Cassava in Food, Feed, and Industry

Authors

C. Balagopalan, Ph.D.
Head
Division of Postharvest Technology
Central Tuber Crops Research Institute
Trivandrum, India

G. Padmaja, Ph.D.
Scientist
Division of Postharvest Technology
Central Tuber Crops Research Institute
Trivandrum, India

S. K. Nanda, M.Tech.
Scientist
Division of Postharvest Technology
Central Tuber Crops Research Institute
Trivandrum, India

S. N. Moorthy, Ph.D.
Scientist
Division of Postharvest Technology
Central Tuber Crops Research Institute
Trivandrum, India

CRC Press
Taylor & Francis Group
Boca Raton London New York

CRC Press is an imprint of the
Taylor & Francis Group, an **informa** business

First published 1988 by CRC Press
Taylor & Francis Group
6000 Broken Sound Parkway NW, Suite 300
Boca Raton, FL 33487-2742

Reissued 2018 by CRC Press

Library of Congress Cataloging-in-Publication Data

Cassava in food, feed, and industry.

Bibliography: p
Includes index.
1. Cassava. I. Balagopalan, C.
TP416.T3C36 1988   633.6'82   88-2626
ISBN 0-8493-4560-X

A Library of Congress record exists under LC control number: 88002626

ISBN 13: 978-1-315-89133-0 (hbk)
ISBN 13: 978-1-351-07043-0 (ebk)

Visit the Taylor & Francis Web site at http://www.taylorandfrancis.com and the
CRC Press Web site at http://www.crcpress.com

PREFACE

Cassava *(Manihot esculenta* Crantz) is the staple food of more than 300 million people in the world. Though cassava is utilized in a variety of ways, scientific books of any category written on the postharvest aspects of cassava are relatively few. The effect of this paucity was strikingly felt during recent years. This was one of the impelling reasons behind the present venture which, it is hoped, will stimulate other publications on this neglected crop.

We have made an earnest attempt to collect information on the origin, spread, and distribution of cassava spoilage and preservation, storage of processed cassava, processing technology, nutritive and toxic factors, physical and chemical nature of starch, utilization in feed, food, industry, and analytical methods for cassava.

Though there are many admirable publications on the storage, processing, and utilization of cassava, no concerted attempt has been made in the past to bring all these aspects under one umbrella confined to the standard befitting the researchers, technocrats, industrialists, students, and planners.

In the writing of this volume, *Cassava in Food, Feed, and Industry,* many recent works on the subject have been consulted. It is neither possible nor necessary to mention them all here but our indebtedness to them is enormous.

We wish to record our thanks to the Editor of A.O.A.C., the Copyright Administrator, American Chemical Society Publishers, Starke Verlag, West Germany, Elsevier, U. K., and the IDRC, Ottawa, for the permission accorded to reproduce some of the methods, tables, and figures in this book. Our thanks are also due to Dr. S. P. Ghosh, Director, Central Tuber Crops Research Institute, Trivandrum, for giving permission to publish this volume and also to use various photographs. Our thanks are especially due to Dr. S. F. Rajiva, Montreal, Canada for the correction of the manuscript and Dr. N. Hrishi, former Director at CTCRI, Trivandrum for his guidance. We are indebted to our colleagues in the Division of Postharvest Technology, CTCRI, Trivandrum for the timely help given during the preparation of the manuscript, and also to the Thribhuvana Typing Center, Trivandrum, for typing the manuscript.

Finally, we are thankful to the Indian Council of Agricultural Research for the inspiration and encouragement given to the researchers for making an endeavor such as this possible.

C. Balagopalan
G. Padmaja
S. K. Nanda
S. N. Moorthy
Trivandrum, India

THE AUTHORS

Cherukat Balagopalan, Ph.D., is Head of the Division of Postharvest Technology, Central Tuber Crops Research Institute, Trivandrum, Kerala, India, under the Indian Council of Agricultural Research.

Dr. Balagopalan received his B.Sc. (Agriculture) and M.Sc. (Agriculture) degrees from Kerala University, in 1963 and 1968, respectively. He received his Ph.D. Degree in Agricultural Microbiology in 1973 from the University of Agricultural Sciences, Bangalore, India. After serving as a research associate at the Kerala Agricultural University, he joined the Indian Council of Agricultural Research as a scientist at the Central Tuber Crops Research Institute in 1974. He has headed the Department of Postharvest Technology of the Institute since 1979.

Dr. Balagopalan was a Visiting Scientist at the Department of Microbiology, University of Guelph, Ontario in 1978 and also at the Department of Biochemical Engineering, Hungarian Food Research Institute, Budapest, Hungary in 1985. The team lead by Dr. Balagopalan won the national award of the Indian Council of Agricultural Research for team research for the biennium 1985 to '86. The award was for the development of an integrated technology for the storage, processing, and utilization of cassava.

Dr. Balagopalan has published more than 50 research papers and also guided students for their Ph.D. degree. His current research interests are recycling of organic wastes for mycoproteins and energy, starch-based fermentations, and utilization of tuber crops.

Gourikkutty Padmaja, Ph.D., is currently working as a scientist in the Division of Postharvest Technology, Central Tuber Crops Research Institute, Trivandrum, Kerala, India, under the Indian Council of Agricultural Research.

Dr. Padmaja received her B.Sc. degree in Chemistry and M.Sc. degree in Biochemistry, in 1972 and 1974, respectively, from the Kerala University, Trivandrum, with distinction. She has taken her Ph.D. on the biochemical aspects of cassava toxicity and is awaiting the award of the degree.

Dr. Padmaja joined the agricultural research service of the Indian Council of Agricultural Research in 1976. She has more than 20 publications to her credit and currently is working on the storage, processing, and utilization of tuber crops. She is also investigating the nutritional and toxicity aspects of cassava in animal systems. Dr. Padmaja has jointly won the National Award of the Indian Council of Agricultural Research for team research for the biennium 1985 to '86 for developing an integrated technology for the storage, processing, and utilization of cassava.

Saroj Kumar Nanda, M.Tech., is currently working as a scientist (agricultural structure and process engineering) in the Division of Postharvest Technology at the Central Tuber Crops Research Institute, Trivandrum, India.

Mr. Nanda received his Bachelor's degree in Agricultural Engineering from Orissa University of Agriculture and Technology, in Bhubaneswar, India, and the University gold medal for securing 1st rank. After completing his M.Tech. at the Indian Institute of Technology, Kharagpur, India, in 1979 he joined the Agricultural Research Service. He has been engaged in research to develop processing techniques and equipment at CTCRI. He has designed approximately six prototypes thus far, including one patented with the National Research and Development Corporation of India. He has 12 publications to his credit.

His current research interests are utilization of renewable sources of energy for processing cassava, development of modern small-scale starch extraction systems, and storage of cassava products.

Mr. Nanda has jointly been the recipient of the ICAR Award for Team Research for the biennium 1985 to '86.

Subramoney Narayana Moorthy, Ph.D., is currently working as a scientist in the Postharvest Technology Division of the Central Tuber Crops Research Institute, Trivandrum, India. After completing his Bachelor's degree in Chemistry in 1969, he obtained his Master's and Doctorate degrees from the Indian Institute of Technology, Kanpur, India, in 1969 and 1975, respectively.

He joined CTCRI as a scientist through the Agricultural Research Service in 1976. Since then he has worked on the chemistry of tuber crops starches, utilization of tuber crops for various food products, and biochemical evaluation of the various tuber crops. He has over 30 publications in national and international journals. He has participated in an international training program on cassava utilization organized by Centro Internacional Colombia de Agricultura Tropical, Colombia, in 1986. Dr. S. N. Moorthy has jointly won the National award of the Indian Council of Agricultural Research for team research for the biennium 1985 to 86, in agriculture, for developing an integrated technology for the storage, processing, and utilization of cassava.

TABLE OF CONTENTS

Chapter 10
Cassava Based Industries

Chapter 1

INTRODUCTION

I. BACKGROUND

Cassava (*Manihot esculenta* Crantz), also known as tapioca or manioc, is one of the major tuber crops grown in more than 80 countries of the humid tropics. It is a high energy food obtained with low inputs and little effort. To the people in the tropics it is either a main or a secondary staple food. The annual production of cassava in the world is 129.02 million t from a net cropped area of 14.150 million ha distributed in over 80 countries.[1] Most of the world production of cassava is used for human consumption in tropical countries, the other main uses are for animal feed and the starch industry.

II. THE CROP AND ITS PRODUCTION

Cassava is a perennial plant growing to a height ranging from 1 to 5 m with a three core single or multitier branching stems. However, variations in branching habit and color of the stems and petioles can be noticed among the cultivars. The leaves are deeply, palmately lobed and the roots are enlarged by deposition of starch cells which constitute the principal source of food. Tuberization occurs usually between the 45th and 60th day after planting and tuber bulking is a continuous process. However, from a quality point of view, the crop is usually harvested between 10 to 12 months after planting. Tubers are usually 5 to 10 in number per plant and develop radially around the base of the stem cuttings by the process of secondary thickening of the xylem parenchyma cells. The tubers are cylindrical or tapering and normally 15 to 100 cm long and 3 to 15 cm in diameter. The tubers consist of an outer skin or periderm which may be white, brown, or pink in color. It has a thin rind or cortex and a core or pith rich in starch. The core is most often white but is sometimes yellow or tinged with red.

An average tuber yield of 5 to 12 t/ha has been reported by traditional methods of cultivation, but by cultivating high yielding varieties and following better package practices the yield can go up to 40 to 60 t/ha. Cassava is one of the most efficient convertors of solar energy. The starch content in its large thickened storage roots may be 25 to 35% on fresh weight basis, but on a dry weight basis, the starch content may go up to 85%. Its productivity in terms of calories per unit land area per unit of time, appears to be significantly higher than other staple food crops. Cassava can produce 250×10^3 cal/ha/day compared to 176 $\times 10^3$ for rice, 110×10^3 for wheat, 200×10^3 for maize, and 114×10^3 for sorghum.[2]

III. CONSUMPTION PATTERN

According to the Food and Agriculture Organization (FAO) food balance sheet for 1975 to 1977, 65% of the total cassava production goes for human consumption and the rest goes for animal feed and industrial uses. The food consumption pattern of cassava significantly varies from different countries. In some countries cassava is fermented before making any food preparation. Direct use of cassava by baking or boiling is the common practice in a majority of the countries. Though cassava provides adequate calories when consumed as a food, there is experimental evidence to show that it imparts several nutritional deficiency symptoms when consumed indiscriminately without adequate protein supplements. The high incidence of protein/calories malnutrition and nutritional diseases among children in the developing countries calls for effective steps to improve the quality and quantity of diets.

It is well known that the most vulnerable group to be adversely affected by nutritional deprivation are pregnant and lactating mothers, and children. Among the various strategies that have evolved to combat malnutrition, enrichment of cassava diets with protein assumes considerable significance.

IV. STORAGE AND PROCESSING

Since cassava constitutes man's staple food in the rural areas of Asia, Africa, and South America, there is a great deal to be done in respect of preserving and storing this perishable commodity. It is also necessary to improve the processing technology of cassava because of the rapid spoilage and lack of storage facilities. Preliminary work undertaken elsewhere in the world has shown that it is possible to prolong the shelf life of cassava tubers. Development of low cost or no cost technology could help the small farmers to store their produce without additional expenditure.

V. CASSAVA STARCH

Starch continues as the major industrial produce from cassava. Though cassava starch possesses a superior property of bland taste, it has not found much use in the food industry. The traditional way of cooking still persists and the adverse property attached to cassava flour and starch is its stickiness in cooking. The cohesiveness can be reduced by cross-linking the starch molecules which will lead to the strengthening of starch granules. Attempts to strengthen the molecules by various simple treatments can be tried so that cassava flour and starch can be increasingly incorporated in various proportions. If the modifying agent carried nutritive value, the fortification of the final food can also be achieved. Pregelatinized starch with proper fortification can be used in infant food and other food items.

VI. COOKING QUALITY

A better understanding of the cooking quality of cassava and the reasons for its variations under varietal and environmental differences can help in increasing the acceptability of cassava for direct consumption as food. However, development of simple and economic processing technology for convenience and instant food can offer greater avenues for utilization of this food crop rich in starch.

VII. CATTLE FEED

In view of the acute shortage of cattle feed, formulations based on cassava have to be developed. Protein-enriched (conventional and nonconventional sources of proteins) cassava-based cattle feed could be formulated at the farm level if farmers are supplied with formulations based on the research on these subjects. Diversification of the uses of cassava could stabilize the price of the produce in the market, which could ensure the adoption of advanced crop production technology (Figure 1).

VIII. NONCONVENTIONAL PRODUCTS

Serious repercussions consequent to the spiraling price-hike of oil resulted in the development of alcohol-driven motor vehicles in many parts of the world. Cassava is being considered as one of the raw materials for the production of alcohol. An enormous quantity of alcohol is also required for the production of a variety of chemicals, such as cellulose triacetate, vinyl acetate, PVC, styrene, and polystyrene. Alcohol from cassava can solve

FIGURE 1. Cassava products.

the alcohol shortage problems in many countries.

Sugar production in the world comes mainly from sugarcane. Dependence on cane sugar can be reduced by introducing the revolutionary sweetner high-fructose syrup with a dextrose equivalent of 95 to 98%. With the development of an appropriate process, cassava starch can serve as a promising alternative substrate for the production of high-fructose syrup.

An attempt has been made in the succeeding chapters to describe the importance of cassava and its application in food, feed, and industry.

REFERENCES

1. Food and Agriculture Organization, *Production Year Book*, Vol. 38, FAO, Rome, 1984, 130.
2. **Coursey, D. G. and Haynes, P. H.,** Root crops and their potential as food in the tropics, *World Crops*, 22, 261, 1970.

Chapter 2

CASSAVA ORIGIN, DISTRIBUTION, AND CULTIVATION

Tuber crops constitute one of the most important groups of crops of tropics and cassava is one of the most widely cultivated tuber crops serving as the major staple food of more than 300 million people in the tropics. Nearly 65% of the total production is utilized for food and the rest is utilized in the feed and industrial sector.

I. BOTANY

Cassava (*Manihot esculenta* Crantz) belongs to the family Euphorbiaceae. It is a perennial woody shrub producing enlarged tuberous roots and variously branched stems. (The leaves are generally large and palmate with 5 to 9 or more deep lobes.) The crop is basically cross-pollinated and hence highly heterozygous. But due to vegetative propagation a high degree of self-pollination does occur in nature.

The height, spread, and other characteristics vary among cultivars. Usually cassava attains a height varying from 1 to 5 m. The shrubs of cassava may be multibranched or unbranched. A large degree of variation in morphology exists within the species of cassava in foliage, stem, and tuber characteristics.

The stem of the cassava plant is tall, thin, and straight and has a number of nodes. The color of the stem usually varies from red-brown to gray. The diameter of the stems varies with age ranging about 3 to 6 cm. Stem branching during growth occurs at different heights namely one third, one half, two thirds, or at the apex of the plant producing primary branchlets in doubles or quadruples which may in turn form secondary and tertiary branches (Figure 1). In the branching types of cassava, branching may occur at the base of the stem, midway up the stem, or at the top. The number of branches vary from 1 to 6 depending on the variety. Cassava is grown by planting cuttings taken from the woody part of the stem. During germination one or more of the auxiliary buds sprout and the roots grow principally from the base of the cuttings. When the main shoot becomes reproductive, the buds at the distal nodes begin to elongate and some develop as shoots. The shoots show strong apical dominance which suppresses the development of side shoots. When the main shoot becomes reproductive and begins to flower, the apical dominance is broken and several of the auxiliary buds immediately below the apex begin to develop, giving the typical branching habit found in the plant. Leaves of cassava are palmate possessing 3 to 11 lobes. The primary leaves of seedlings are unlobed where as secondary leaves are 3-lobed. Leaves are attached to slender petioles at the base of lamina, which is held on a nearly horizontal plane. Leaves possess slender stipulates which cause scars at abscission. Stipules normally form at the point of attachment of petioles. They abscise readily together with the petioles to form typical prominent stem nodes. These nodes contain dormant buds which give rise to new shoots and adventitious roots.

The tuberization process involves essentially the onset of secondary thickening in some of the adventitious roots which are previously fibrous in nature. This starts at the proximal part of the root. Adventitious roots arise from the base of the cutting. These roots later develop into a fibrous root system which are the main feeder root of the plant. Later on some of these fibrous roots begin to swell and become tuberous. Apparently all the fibrous roots are initially active in nutrient absorption, but once they become tuberous, their ability to function in nutrient absorption decreases considerably. Cassava grown from seed develops a taproot system. The radicle of the germinating seeds grows almost vertically downward and develops into the taproot while a few adventitious roots may originate from the subterranean stem portion. It seems that the taproot invariably becomes tuberous and is the first root in the plant to do so usually.

FIGURE 1. A cassava plant.

The growth cycle of a typical cassava crop is close to 1 year. The roots start bulking about 3 months after planting and continue to increase in weight until 9 to 15 months after planting when the crop is usually harvested. During 75% of its growth period it accumulates carbohydrate.

Flowers are born on terminal panicles. Cassava is monoecious and unisexual. Each flower has five united sepals but not petals. Male flowers have ten stems arranged in two whirls of five each. The flower has an ovary mounted on a 10-lobed glandular disc. It has three locules and six ridges. Each locule has a single ovule. The stigma has three lobes which unite to form a single style. In each inflorescence, the female flowers open first, while the male flowers do not open until about 1 week later. Insects are the pollinating agent, mainly honey bees. It requires 3 to 5 months after pollination for the fruits to mature. The mature fruit is a capsule, six narrow longitudinal wings are present. The endocarp is woody and there are three locules each containing a single seed. When the fruit is dry the endocarp splits explosively to release and disperse the seeds.

II. ORIGIN AND SPREAD

Cassava is a native of tropical America. Until recently Brazil was considered to be the

FIGURE 2. A cassava plantation.

place of origin for cassava. But according to the recent reports the lowlands of Southern Mexico, Guatemala, Honduras, Venezuela, and Colombia are also believed to be the place of origin. The Portuguese who colonized South America by 1500 A.D. in these regions developed cassava cultivation in plantation level and carried it to other continents. In Africa, cassava was introduced in the 16th century in and around the Congo river basins. Cassava was introduced to India by the Portuguese in the 17th century in 1840. In Java it was introduced around 1810. In about 1850 it was transported directly from Brazil to Java, Singapore, and Malaysia (Figure 2).

The total world area and production under cassava in 1984 according to FAO estimates is 14.15 million ha and 129.02 million t respectively. Asia alone accounts for 29.47% of world area and 38.75% of the total production of cassava. Though tropical America is the place of origin for cassava, South America produces only 20.82% of the world production from 16.33% of the world cassava area. Africa produces 37% of the world crop. In Asia the average yield varies from 8 t/ha and in Vietnam to over 16 t/ha in India. In Africa the range is 5 to 9 t/ha. In South America the average production is from 10 to 15 t/ha.[1]

III. CULTIVATION

A. Climate

Cassava can adapt itself to a variety of climatic conditions, though a warm and humid climate is preferred. It comes up well between latitudes of 30° North and 30° South. Cassava is grown at altitudes up to 2000 m but can be grown profitably at lower altitudes in areas ranging from humid (more than 2000 mm annual rainfall) to semiarid (500 to 750 mm) conditions. Cassava has shown its hidden potential to withstand drought conditions. But it cannot withstand exposure to frost and growth ceases at temperature below 10°C. In the extreme latitudes where cassava is grown a light frost occasionally occurs which can defoliate the plant. But sprouting begins from the base of the stem when temperatures increase. The optimum temperature range for the growth of cassava is between 18 to 30°C. Cold tolerant varieties have been selected in South America by farmers to grow cassava below 17°C.

Cassava is a drought-tolerant crop and can be grown in areas where there are occasionally prolonged spans of drought. During a drought period the plant regulates the transpiration rate and enters into a dormant state, shedding older leaves and putting forth fewer new leaves. With the onset of rains, reserve carbohydrates from the roots and stem are utilized and the plant becomes active again. But flooding and excessive moisture in the soil will enhance the rotting of tubers and subsequent spoilage. Cassava is basically adpated to a tropical environment and yield well under full sunlight when the soil moisture is not limiting. It has been established that the mechanism of photosynthesis in cassava follows a C_3 pathway which always inhibits light saturation. The light intensity for maximum photosynthesis for C_3 plants is found to be about 44000 lx which is easily achieved under a tropical climate throughout the year.

B. Soils

Cassava can tolerate a wide range of soil conditions. It grows well in soils of low fertility status and produces a satisfactory yield. A well-drained loamy or sandy loams with sufficient organic matter is the best soil for the cultivation of cassava. Proper tuber formation takes place only in those soils which are loose and friable. On heavy clay and rocky soils, the yield will be poor due to restricted tuber development. Poor drainage and water logging are also unfavorable for growing cassava. The optimum pH of the soil for cultivation is from 6 to 7.5. But it is grown under a wide-range of pH in many parts of the world. In Malaysia cassava is grown in limed peat soils having a very low pH of 3.2. In Kerala (India) the cultivation is mainly located in the laterite and red soils which are acidic in nature, low in cation exchange capacity and base saturation, and high in phosphorus fixation capacity. A high concentration of salts or high soil pH may severely effect the growth and productivity of cassava. Yet it grows remarkably well in soils of low pH and high aluminum levels.

C. Preparation of Land

Under rain-fed conditions, the best time of planting is April to May with the onset of the southwest monsoon. Planting in August to September is also a usual practice. If irrigation facilities are available cassava can be grown in any part of the year.

Loosening of the soil to a depth of 20 to 25 cm either by tractor ploughing or spade digging facilitates better rooting. Based on the type and topography of the soil the method of cultivation also varies. The mound method is adopted in soils having high clay content and restricted drainage, whereas the ridge method is followed in slopes to prevent soil erosion. In level lands having good drainage a flat method of cultivation is followed. Pits followed by mounds is found to be superior over the flat method. Pits of size 30 cm × 30 cm × 25 cm are opened first and the soil is mixed-up with cattle manure and reformed into a mound. Fertilizers have to be applied in a furrow opened at 15 cm radius from the top of the mound and covered with soil. This method of cultivation consumes more labor, but a yield increase of nearly 10% over the conventional method of mound preparation has been recorded.

D. Selection of Planting Materials

Disease and pest-free planting materials of 8 to 12 months maturity having a thickness of 2 to 3 cm may be selected for planting. About a 10 cm portion from the base of the stem and one third length from the top immature portion may be removed and the remaining portion of the stem can be used for preparing stakes. Bottom stakes give higher yields when compared to middle and top portion stakes. A stake length of 20 cm and a planting depth of 5 cm are optimum. The stakes may be planted vertically at 90° to the ground.[2] A spacing of 90 cm × 90 cm is ideal for planting cassava. Replacing the unsuccessful stakes at 15 days after planting with fresh cuttings is advantageous as compared to gap filling at a later date. Removal of excess sprouts by nipping at the initial stages of establishment (10 to 15 days after sprouting) helps to prevent mutual shading and competition between plants.

E. Manures and Manuring

Cassava responds well to the application of organic and inorganic manures. A basal application of farmyard manure at the rate of 12.5 t/ha is required to improve the physical conditions of soil. Use of 100 kg of nitrogen per hectare was found to be the optimum in laterite soils. Application of this dose in two splits, namely one half as basal and the other half after 1 or 2 months after planting was found to be superior to other split applications. Application of urea as one half soil + one half foliar was found to be more effective and economical, and it was found to reduce the hydrogen cyanide (HCN) content of tubers.

In an acid laterite soil, cassava responded to an application of phosphorus up to 100 kg P_2O_5. However, continuous application at this level has resulted in a buildup of available phosphorus status in acid laterite soils and in such soils there is no response to phosphorus application for at least three seasons. Full basal application of phosphorus was superior to other applications and its application at various depths namely surface, 3 and 6 in. depth had no significant influence on the yield of cassava.

Increasing levels of potash application has resulted in an increase in starch content and a decrease in HCN content of tubers. Potassium at 100 kg/ha applied in two split doses namely one half dose as basal plus one half dose, 1 to 2 months after planting was found to be optimum. The tuber yield was found to decrease beyond 100 kg K_2O/ha, whereas the concentration of potassium in the plant increased which might be attributed to the luxury consumption of potassium by the plant.

Experiments have shown that the potash requirement of cassava is as high as nitrogen. Nitrogen and potash at 100 kg/ha (in the ratio of 1:1) was found to be optimum for tuber yield. Response of cassava to incremental doses of N and K_2O was worked out and it was found that as the doses increased, the unit increase in yield per kilogram of nutrient decreased.[3]

Application of 2000 kg CaO/ha was found to be beneficial in increasing the yield and quality of cassava in India. This has increased the soil pH by two units (4.5 to 6.5) and increased the available P and exchangeable Ca content of the soil, thereby increasing their uptake.

Application of sulfur at 50 kg/ha gave significantly more tuber yield of cassava (18% more than control) and decreased the starch and HCN content of tubers. The methionine and total protein content of tubers were also found to increase by the application of sulfur.

Application of zinc has gained considerable importance recently because of the higher yield of tuber after zinc application. Zinc, boron, and molybdenum at 12.5, 10.0, and 1.0 kg/ha respectively, as zinc sulfate, borax, and ammonium molybdate have increased the yield of cassava in laterite soils.

F. Intercultivation and Earthing-Up

In order to control weeds and improve the soil conditions interculturing is essential. Soil has to be occasionally stirred and heaped around the base of the plant. The first interculturing and earthing-up may be given 45 to 60 days after planting, and the second interculturing and earthing-up 1 month later. Weeds compete with cassava, especially in the early stages of its growth for nutrients. Hence, they have to be controlled for the first 3 to 4 months after planting. Besides hand-weeding and interculturing, commercial chemicals can be also used for weed control in the initial stages of growth. Diuron can be used against annual weeds and Alachlor against grasses. A mixture of both the chemicals can also be tried.

G. Irrigation

Cassava is normally grown as a rain-fed crop but responds well to irrigation. Irrigation at 25% available moisture depletion level throughout the growing season could double the tuber yield as compared to control. As a result of irrigation, the starch content could be increased and HCN content of irrigated crops could be reduced.

H. Multiple Cropping Systems

A multiple cropping system is one of the most intensive systems in crop management and is one of the most productive and useful approaches for the small and subsistence farmers. Multiple cropping includes different cropping systems such as intercropping, mixed cropping, multistoried cropping, sequential cropping, and relay cropping. This system provides the farmer with a variety of returns from the land, often increases the efficiency of resource utilization by combining variety of crops, and reduces the risk of dependence on a single crop which may suffer from environmental or economical fluctuations. It is also one of the methods that reduces soil erosion, leaching of nutrients, depletion of fertility, and rapid growth of weeds.

Intercropping with grain legumes like cowpea (*Vigna unguiculata*), mung-bean (*Phaseolus aureus*), blackgram (*P. mungo*), soy bean (*Glycine max*), and Pigeon pea (*Cajanus Cajan*), oil seeds such as groundnut (*Arachis hypogaea*), sunflower (*Helianthus annus*), and vegetables such as cucumber (*Cucumis sativus*), french bean (*P. vulgaris*), cowpea (*V. sinensis*) and bhindi or okra (*Abelmoschus esculenta*) were found successful. Among these crops cowpea, pigeon pea, groundnut, and french beans were found to be very economical. Corn can be also grown as an intercrop on cassava plantations.

Intercropping of cassava in coconut, arecanut gardens, and banana plantations is a usual practice in the state of Kerala in India. Since cassava is known to be a sun-loving plant and requires abundant sunshine for maximum production, the yield of cassava under shaded conditions is very poor and ranges from 6 to 8 t/ha. But yield will be better in coconut and arecanut plantations when cassava is grown during the first few years of the plantation.

Cassava can also be grown as a sequential crop in lowlands where there is sufficient moisture available after the harvest of a rice crop. Early maturing varieties are preferred in lowlands. The crop can be harvested 6 to 8 months after planting.

I. Diseases and Pests

1. Mosaic Disease

The African mosaic disease or cassava mosaic disease causes heavy yield losses in Africa and India. This disease is not present in South and Central America. Early development of leaf chlorotic areas can be observed and leaflets are distorted. As the chlorotic areas, bright yellow or pale green in color separated by normal tissue, enlarge it gives rise to a typical mosaic pattern. In the susceptible varieties the leaves are reduced in size, misshapen, and twisted giving a leaf-curl appearance. Indiscriminate use of diseased stems is one of the possible routes for the spread of the disease in the succeeding generations. The white fly *Bemisia tabaci* transmits the disease in the field. As the disease spreads mainly through use of infected sets, use of disease-free stems as planting material can considerably reduce the incidence of the disease.

2. Brown Leaf Spot

Due to the infection of leaves by *Cercospora henningsii*, severe premature defoliation resulting in great reduction of photosynthetic area and subsequent reduction in yield is the aftereffect of the infection. The disease causes characteristic brown spots (3 to 12 mm in diameter) on both sides of the mature leaves. As the disease develops the infected leaves turn yellow, dry out, and fall off. Benlate (0.1%) and cercobin (0.2%) significantly reduced the disease intensity and premature leaf fall, and increased the tuber yield when sprayed six times at monthly intervals starting from the 3rd month onwards.

3. Cassava Bacterial Blight

This disease is widely distributed in South and Central America, Asia, and Africa. Gray, angular leaf spots that develop into large brown areas with a water soaked appearance are typical symptoms of the disease. The disease is severe when rainfall is high and fluctuations in temperature are experienced. Resistant varieties have been evolved at IITA Nigeria and CIAT Colombia against bacterial blight.

4. Superelongation

This is a disease widely present in the Central and South America. Symptoms are a marked elongation of the internodes and a malformation of the leaves. Infected cuttings and spores spread the disease over long distances. Resistant varieties have been identified at CIAT, Cali, Colombia.

5. Set Rot

Rotting of sets caused by *Diplodia natalensis* results in the failure of germination and gap formation in the field. Dipping the stems in fungicides such as thiram (0.2%), benylate (0.2%), bavistin (0.1%), and vitavax (0.1%) were found to be effective in checking the rot.

6. Dieback

The dieback disease of cassava is caused by *Glomerella cingulata*. Young shoots and growing tips are more susceptible. The tissues become necrotic and die due to this disease. When the stem is attacked they die causing a dieback of young stems. One or two spraying of dithane M-45 (0.3%), thiram (0.3%), or dithane (0.1%) would control the disease.

7. Red Spider Mites and Mealy Bugs

These are the most destructive insect pests of cassava. The spider mites begin to feed first on the lower leaves producing characteristic yellow to reddish-brown spots on the upper surface of the leaves along the central veins. Eventually the symptom spreads over the whole leaf which turns reddish-brown in color. The affected plants become severely stunted and yields are considerably reduced. Heavy infestation occurs in summer. Spraying the crop 3 times with dimethonete or thiocron at 0.3% concentration can control the disease.

8. Thrips

They lower yields by destroying and reducing the size of new leaves, stunting the new growth of the plant, and damaging the apical meristem.

Termites, cockchafers, stem borers, leaf-eating grasshoppers, scale insects, and horn worms are some other pests which attack cassava in different proportions.

9. Rats

Rats are a serious menace to the crop as they destroy on an average about 5 to 10% of the total production of fresh tubers. They burrow into the soil and feed on the tubers. The use of traditional traps and rat poisons are the only method to stop the rat menace.

J. Harvesting

Cassava plants are normally ready for harvest from the 7th month on, and harvesting can continue until 18 months after planting or later. The starch content of majority of varieties reached a peak of 10 to 11 months. If harvesting is delayed the tubers become fibrous. The common method of harvesting cassava is by hand. In hard soils the stem is cut first leaving a few corms above the ground. This is pulled out along with the tubers. In order to reduce the harvesting time and work load, a simple harvesting aid has been developed at IITA, Nigeria. A tractor drawn device designed at CIAT, Colombia loosers the roots and leaves them on or close to the soil surface. The roots are then picked up by hand, separated from the stem, and packed for transport. Completely mechanical devices have been developed for the harvest and direct loading of the tubers for the purpose of transport.

REFERENCES

1. **Cock, J. H.**, Cassava plant and its importance, in *Cassava New Potential for a Neglected Crop*, Westview Press, Boulder, Colo., 1985, chap. 1.
2. **Nair, G. M., Ravindran, C. S., and Mangal, P.**, Cultural practices for tuber crops, *Indian Farming*, 33, 29, 1984.
3. **Kumar, M. B., Kabeerathumma, S., and Nair, P. G.**, Soil fertility management of tuber crops, *Indian Farming*, 33, 35, 1984.

Chapter 3

CASSAVA NUTRITION AND TOXICITY

I. NUTRITIVE VALUE OF CASSAVA

The calculated nutritive value of cassava has been reported to be higher than that of corn or sorghum. Although a number of reports are available on the detailed chemical composition of cassava, the results vary widely depending on the age of the plant, variety, climatic conditions, cultural practices followed, etc.[1-6] The roots and the leaves of the cassava plant are the two nutritionally valuable parts which offer potential as a food or feed source. The principal parts of the mature cassava plant expressed as a percentage of the whole plant are leaves 6%, stem 44%, and tubers 50% (comprising 11% water, 8% peelings, and 31% starch).

A. Cassava Roots: Nutritive Value

The cassava root has an average composition of 60 to 65% moisture, 30 to 35% carbohydrate, 0.2 to 0.6% ether extractives, 1 to 2% crude protein, and a comparatively low content of vitamins and minerals.[7] However, the roots are rich in calcium and vitamin C and contain a nutritionally significant quantity of thiamine, riboflavin, and nicotinic acid. The carbohydrate fraction contains 3.2 to 4.5% crude fiber and 95 to 97% nitrogen-free extract (NFE).[8,9] The tuber NFE contains 80% starch and 20% sugars and amides.[10] The starch content increases with the growth of the tubers and reaches a maximum between the 8th and 12th month after planting. Thereafter the starch decreases and the fiber content increases. According to Johnson and Raymond,[2] cassava starch contains 20% amylose and 70% amylopectin. Cassava root meal has been reported to have about one third the amylolytic activity of corn and is highly digestible, yielding a digestible energy of 400 kcal/kg of dry root meal for pigs as compared with 4,055 kcal/kg of maize.[9] Cassava root also contains sucrose, maltose, glucose, and fructose to limited levels.[11] Sreeramamurthy[12] studied the digestibility of cassava starch and it was reported that the raw starch has a digestibility of 48.3%, while cooked starch has a digestibility of 77.9%. Cassava starch was digested to a greater extent by "taka" diastase than by pancreatic amylase.[13]

Cassava root is a poor source of protein making the economic utilization of roots in animal feeds highly dependent on the incorporation of other protein-rich ingredients (Table 1).[14] The quality of cassava root protein is fairly good as far as the proportion of essential amino acids as a percentage of total nitrogen is concerned.[6,8,9,14,15] However, methionine, cysteine, and cystine are limiting amino acids in the roots. Only about 60% of the total nitrogen is derived from amino acids and about 1% of it is in the form of nitrates, nitrites, and HCN. The remaining 38 to 40% of the total nitrogen remains unidentified.[16] Maner[15] has reported that out of the total nitrogen, only 50% existed as true protein while the remaining existed in the form of free aspartic and glutamic acids. In contrast to this, Oyenuga[1] reported that 62% of the total nitrogen was true protein. The levels of total nitrogen and nonprotein nitrogen are higher in the bark than in the whole root.[13] The amino acid levels of cassava roots are given in Table 2. The levels of lysine and tryptophan are high in the true protein fraction (Table 2).[14]

According to Hudson and Ogunsua[17] cassava flour contains about 2.5% lipids. Of this, only 50% is extractable with conventional solvents. The extractable lipids are mainly polar in nature, the principal group being galactosyl diglycerides. The component fatty acids are relatively saturated compared with the structural lipids of the potato.

Although a relatively poor source of minerals and vitamins, there is a high content of

Table 1

PROXIMATE ANALYSIS OF CASSAVA AND SOME OF ITS PRODUCTS[14]

	Cassava tubers	Cassava flour	Cassava macaroni	Gari	Fufu
Moisture (g/100 g)	59.4	9.50	10.60	14.40	15.30
Protein (g/100 g)	0.70	1.60	11.20	0.90	0.60
Fat (g/100 g)	0.20	0.40	1.90	0.10	0.14
Crude fiber (g/100 g)	0.60	0.80	0.70	0.40	0.20
Carbohydrate (g/100 g)	38.10	84.90	73.80	81.80	75.80
Ash (g/100 g)	1.00	1.80	1.80	1.40	0.50
Calcium (mg/100 g)	50.00	60.00	30.00	70.00	160.00
Phosphorus (mg/100 g)	40.00	80.00	140.00	40.00	20.00
Thiamine (mg/100 g)	0.05	0.08	0.22	—	—
Iron (mg/100 g)	0.90	3.50	2.90	2.20	6.20
Vitamin C (mg/100 g)	25.20	—	—	—	—
Calories (kcal/100 g)	157.00	338	351	323	393

Table 2

ESSENTIAL AMINO ACID PROFILE (%) ON DRY WEIGHT BASIS[14]

	Free amino acids		Amino acid in protein	
	Tuber	Leaf	Tuber	Leaf
Arginine	0.29	1.48	7.74	5.21
Histidine	0.27	0.66	1.50	2.47
Isoleucine	0.03	1.67	5.33	4.12
Leucine	0.31	2.72	5.56	10.00
Lysine	0.07	1.87	6.23	7.11
Methionine	0.03	0.36	0.60	1.45
Phenylalanine	0.03	0.92	3.45	3.87
Threonine	0.03	1.35	3.83	4.70
Tryptophan	—	0.24	0.53	1.09
Valine	0.04	0.99	4.51	6.18

calcium and phosphorus in the tubers. Barrios and Bressani[18] reported that the mineral content of the dry bark is higher than that of the cortex. Calcium values in the whole root range from 15 to 129 mg/100 g, while phosphorus values are almost constant throughout the various parts of the root, approximately 100 mg/100 g. The content of iron in the central cylinder is 32 mg/100 g, while in the bark it is 77 mg/100 g.

Rojas[19] determined the vitamin C content of six cassava clones from Peru and found that

Table 3
NUTRITIVE VALUE OF CASSAVA LEAVES COMPARED WITH SOME COMMON FODDER CROPS[25]

	Proximate analysis (g/100 g dry matter)				
Fodder crops	Crude protein	Crude fiber	Nitrogen-free extract	Ether extractives	Total ash
Cassava leaves (*Manihot esculenta* Crantz)	23.0	24.4	36.2	4.8	11.6
Guinea grass (*Panicum maximum*)	7.7	37.3	39.4	1.7	13.9
Elephant grass (*Pennisetum purpureum*)	6.2	28.1	47.5	2.3	16.0
Lucerne (*Medicago sativa*)	20.2	30.1	36.6	2.3	10.7
Sweet potato (*Ipomoea batatas*)	17.2	19.3	43.4	3.4	16.7
Leucaena (*Leucaena leucocephala*)	16.7	12.6	51.1	7.1	12.5

the raw roots contained 38.5 to 64.6 mg, while fried and boiled samples contained 29.1 to 47.8 mg and 21.5 to 40.6 mg, respectively. However, drying reduces the vitamin C content appreciably, with the values going down up to 2.8 to 13.0 mg.

B. Cassava Leaves: Nutritive Value
The annual yield of cassava foliage has been reported to be as high as 20 t/ha.[20] Dried cassava leaves are a good source of proteins, minerals, and vitamins. Hence, the leaves offer vast scope as a protein ingredient in compounded feeds for livestock and poultry. Bangham,[21] Echandi,[22] Juarez,[23] and Miranda et al.[24] were among the first to suggest the value of cassava leaves as animal feed. The proximate analysis of cassava leaves compared with some of the common livestock forages are given in Table 3.[25] Echandi[22] found that the protein content of cassava foliage meal (leaves and stems) was similar to that of alfalfa (16.9 to 17.0%). Cassava leaves are superior to many other leaves as there is less crude fiber and a high concentration of fats and carbohydrates. Juarez[23] determined the leaf production of 16 cassava varieties and found that the total yield of leaves could be increased by cutting at 7 and 11.5 months. Paula and Cavalcanti[26] analyzed the amino acid content of cassava leaves and stressed the use of leaves as animal feed. Cassava leaf blades are especially rich in protein (average 30.5%) and the protein value reduces to 13.0% for the whole foliage.[27]

Pechnik et al.[28] determined the amino acids of cassava leaves and found that essential amino acids such as phenylalanine, valine, tryptophan, and arginine as well as nonessential amino acids such as aspartic acid, alanine, proline, glutamic acid, serine etc., were present in fairly good amounts in the leaves. Cassava leaves and roots are low in methionine with values of 1.7 and 1.2 g/100 g of crude protein compared with 2.2 for the FAO reference pattern. Lysine content is high in the leaves (7.2) and low in the roots (3.90) compared with FAO pattern (4.2). Luyken et al.[29] showed that tryptophan, methionine, and cystine levels were low in cassava leaves. The biological value of cassava foliage is variable and is inferior to that of animal protein.

Cassava foliage protein can be produced by drying the leaves in the sun for 2 days on a concrete floor at a density of 15 kg/m². Industrial protein extraction is expensive and uneconomical.

The fresh cassava leaves have a carbohydrate content of 11.4%[18] while dry leaves have

46.5% carbohydrate.[18,27,30] The dry cassava foliage has a carbohydrate content of 43.5%.[22,27,31] A major proportion of the leaf carbohydrate is starch. The amylose content of the leaf starch has been reported to vary from 19 to 24%.[13]

As compared to other tropical forage species, the crude fiber content of cassava leaves is low which makes it palatable as a poultry feed. The leaves are rich in calcium but low in phosphorus compared to corn and sorghum. Pechnik et al.[28] studied the effect of boiling on the mineral content of leaf flour and found that the raw leaf flour and boiled leaf flour contained respectively 1,120 and 1,070 mg% Ca, 300 and 250 mg% P, and 31 and 30 mg% Fe.

Cassava foliage meal has been reported to contain as high as 56,000 IU of vitamin A as compared with 14200 IU in alfalfa meal, 66 IU in ground yellow maize, and 264 IU in wheat flour.[32] This high content of vitamin A is significant in the pigmentation of egg yolks.

The economics of cassava foliage production has been studied by many workers. Scholz[33] obtained 3 t of dry cassava leaves by pruning the plants at 9 months. Meyreles and McLeod[34] planted cassava at different densities to find out the optimum conditions for maximal foliage yield. They found that the best density was 60,000 plants/ha and the foliage should be cut from the age of 3 months at every 3 months. Montaldo and Montilla[35] obtained a foliage yield of 31.86 t/ha (equivalent to a crude protein yield of 5.8 t/ha). In addition to this, this much foliage can also yield 13.8 t of total carbohydrates. Cassava foliage is thus a highly nutritive and economically feasible high protein ingredient of animal feed rations.

II. TOXIC PRINCIPLES IN CASSAVA

From very early days it had been realized that cassava contained toxic principles. The first reference to this noxious principle was made by Clusius[36] and the association of toxicity with hydrocyanic acid was first made by Henry and Boutron-Charland.[37] The presence of HCN in sweet and bitter cassava cultivars was established by Francis[38] and further investigated by Carmody,[39] Collens,[40] and Turnock.[41] In 1886, Peckolt identified the toxic principle, which was originally named mannihotoxin as a cyanogenic glycoside.[42] This compound was later shown[43] to be identical with the glycosides of flax (*Linum usitatissimum*[44]) and from the beans (*Phaseolus limensis*). For this reason this was named linamarin. Although it was believed for several years that this was the only glycoside of cassava, it was later confirmed that another glucoside lotaustralin, which is a higher homologue of linamarin was also present in cassava[46-48] (Figure 1).

The two glycosides, linamarin and lotaustralin are present in all parts of the cassava plant. The synthesis of the glycosides and the storage in various organs take place throughout the life cycle of the plant.[47] The concentration of the glycosides vary considerably between varieties and also with climatic and cultural conditions. The normal range of cyanoglucoside content is from 15 to 400 ppm calculated as mg HCN/kg fresh weight but occasionally varieties with very low HCN content 10 mg/kg or very high HCN content 2000 mg/kg have also been encountered.

The capacity of *Manihot esculenta* to produce HCN varies in the tissues of the same clone or different clones. The distribution of cyanogenic glycoside in different parts of four clones was studied by De Bruijn.[49] She found that the glucoside concentration of the barks of tuberous roots and leaves of less toxic clones was only slightly lower than those of the highly toxic clones. There are also many variations in the glucoside concentration within a tuberous root. In their studies, Bourdoux et al.[50] found that there did not exist an exact correlation between the cyanide content in the tubers and the part sampled for analysis.

Cassava is often classified as "bitter" or "sweet" according to the amount of cyanide present in it. However, several studies have shown that bitterness or sweetness could not be exactly correlated with the level of cyanogenic glycosides.[51,52] However, the cyanide

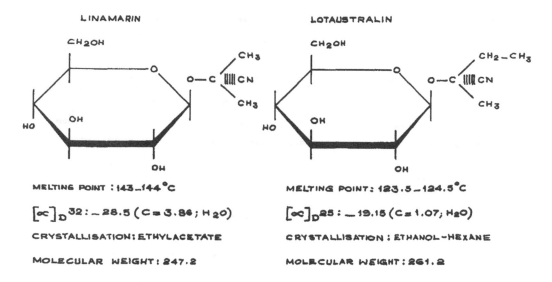

FIGURE 1.   Structure and properties of cassava cyanogenic glycosides.[50]

production is confined to the outer peel generally where a high cyanide gradient exists. The parenchymal tissue has a low content of cyanide.[47]

Based on the amount of cyanogenic glucoside, a tentative classification has been made as[51,53,54]

1.   Innocuous — less than 50 mg HCN/kg fresh peeled tuber.
2.   Moderately poisonous — 50 to 100 mg HCN/kg fresh peeled tuber.
3.   Dangerously poisonous — over 100 mg HCN/kg fresh peeled tuber.

However, there is no definite correlation between the content of cyanogenic glycosides and bitterness.

A. Cyanogenic Glycosides: Biosynthesis

Since many cyanophoric plants and plant products are used as human food or animal feed, there is a great deal of interest recently in understanding the metabolism of cyanogenic glycosides. The toxicity attributed to these foodstuffs is due to the release of hydrogen cyanide on ingestion. Using labeled precursors the biosynthetic pathway of many cyanoglucosides could be worked out.

The biosynthetic pathway of linamarin and lotaustralin was first investigated by Bediako et al.[55] using labeled volatile precursors of linamarin like isobutyronitrile, isobutyraldoxime, and 2-hydroxy isobutyronitrile. The enzymatic pathway was found to be analogous to the scheme demonstrated in *Linum* sp. and *Sorghum bicolor.* The scheme for the biosynthesis of linamarin and lotaustralin is demonstrated in Figure 2. The precursor amino acids for linamarin and lotaustralin are L-valine and L-isoleucine respectively.

From the data on the incorporation of ^{14}C from valine into linamarin, it was concluded that the petioles were the major synthetic site of linamarin in cassava followed by the leaves, upper stem, and shoot apex. The roots and the lower stem had minimal ability to incorporate the label. Ringing experiments by De Bruijn[49] could demonstrate the accumulation of cyanogenic glycosides above the point of incision. This indicated the possibility of cyanoglycosides being synthesized in the upper parts of plant and then translocated to the lower parts. Labeled experiments by Bediako et al.[55] also showed the translocation of synthesized cyanogenic glycosides down the cassava plant from the leaves.

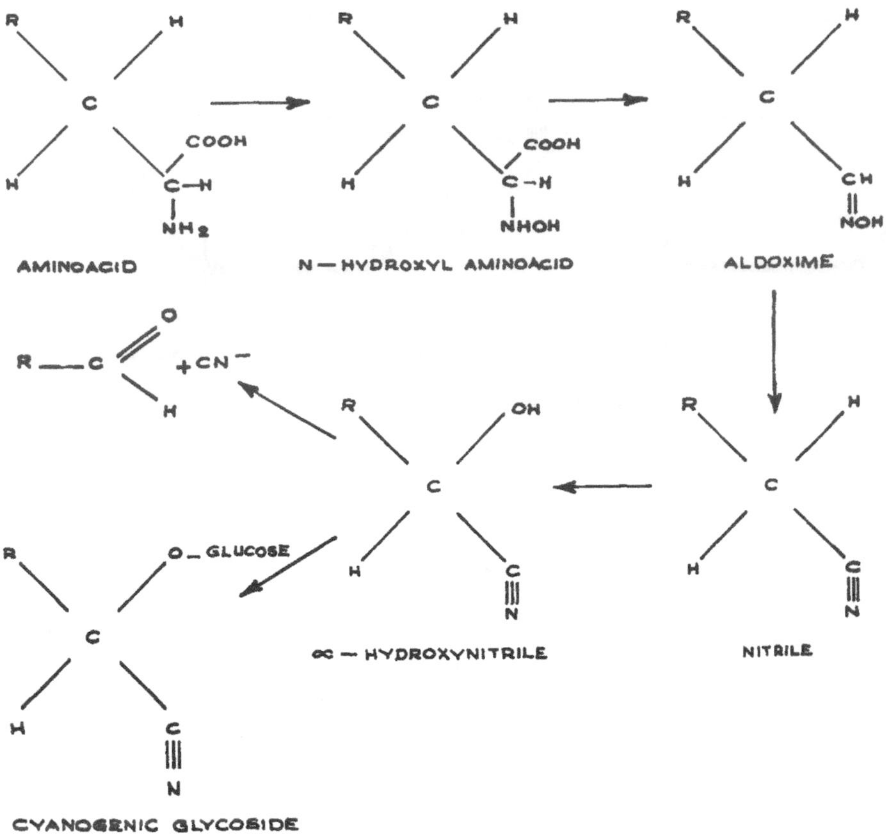

FIGURE 2.   Biosynthetic scheme for cyanogenic glycosides.[60]

B. Cyanogenic Glycosides: Catabolism

It is well recognized that the plants containing a cyanogenic glycoside also contain enzymes capable of degrading it although some exceptions have been reported.[56,57] The degradative pathway has been studied in detail in other plants like *Sorghum bicolor*.[58,59] The cassava cyanogenic glycosides are also believed to follow the same catabolic pathway.[60] The scheme for catabolism of linamarin is as shown in Figure 3.

Plant glycosidases hydrolyzing the cyanogenic glycosides were found to have high specificity towards the cyanogenic glycosides.[61] Linamarase hydrolyzing linamarin also belongs to the group of β-glycosidases. The cassava enzyme is highly specific.[62] De Bruijn[49] found that the activity of linamarase was highest in the young expanding leaves and lowest in the inner part of tuberous roots and the lower part of the stem.

The preparation of linamarase was described from cassava peel using acetone precipiation by Wood.[63] The extraction of the enzyme from cassava parenchymal tissue was described by Cooke et al.[64] Extensive purification of the enzyme could be achieved by using DEAE cellulose chromatography.

There exists an excellent compartmentation in the plant between the cyanogenic glycoside and the degrading enzymes. Thus, in the healthy plant these two will not come together and as soon as the tissue is wounded, the enzymes come into contact with the substrate-liberating free-hydrocyanic acid.

C. Physiological and Biochemical Effects of Cyanide and Cyanogenic Glycosides

Ingestion of cassava can trigger several toxic manifestations due to the release of HCN

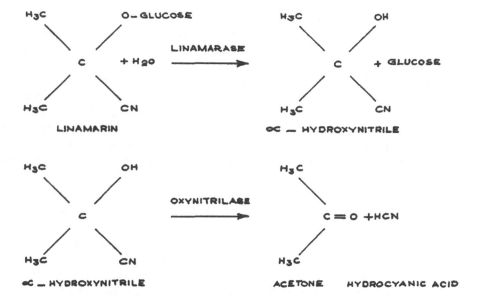

FIGURE 3.   Enzymatic degradation of cassava cyanogenic glycosides.[60]

from cassava cyanogenic glycosides. The incidence of acute poisoning from consumption of cassava is relatively rare since the amount ingested is often low. However, chronic intake of cassava can lead to toxic conditions as the person is exposed to sublethal doses of cyanide for a prolonged period. The toxicity of cassava is due to the release of HCN in vivo which is a potent cytotoxin exerting a wide range of biological effects. The inhibitory properties of cyanide on metabolism have been studied in detail by Solomonson.[65]

The most important effect of cyanide is the inhibition of tissue respiration as it is a potent inhibitor of terminal oxidase of the mitochondrial respiratory chain. A number of other enzymes like catalase, superoxide dismutase, nitrate reductase, etc. are also inhibited by cyanide.

Cyanide inhibits the terminal cytochrome oxidase by combining with it. The nature of interaction between cyanide and cytochrome oxidase has been studied in detail. Cyanide lethality could occur without apparent inhibition of liver cytochrome oxidase.[65] From this they concluded that brain cytochrome oxidase may be the site of lethal action of cyanide. Earlier studies also demonstrated the inhibition of brain cytochrome oxidase by cyanide.[65]

The effect of sublethal doses of cyanide on glucose catabolism was studied in mice by Isom et al.[66] They found that cyanide caused a 100% increase in the catabolism of glucose via the pentose phosphate pathway and a 50% decrease in its breakdown via the glycolytic pathway. Increased blood glucose and lactic acid levels were reported in animals given sublethal doses of cyanide. A gradual shift from aerobic to anaerobic metabolism resulting from a decrease in ATP/ADP ratio has also been reported.[67] Studies with isolated perfused rat livers treated with cyanide showed increased glucose production and phosphorylase activity.

Cyanide inhibits a number of metalloenzymes such as alkaline phosphatase and cytosolic Cu/Zn superoxide dismutase. It also forms a 1:1 inhibition complex with carbonic anhydrase. Glutathione peroxidase is also inactivated by cyanide by the removal of selenium from it. A number of enzymes are also inactivated by cyanide through nonmetallic interactions. Rat brain glutamate decarboxylase and γ-aminobutyric acid transaminase are likewise inhibited by cyanide.[65]

1. Acute and Chronic Effects of Cyanide

Ingestion of cyanide-containing staple foods such as cassava and forage plants have accounted for many instances of cyanide poisoning in humans and animals. The clinical signs and symptoms of cyanide poisoning were described over 150 years ago. The cyanide ion is rapidly absorbed from the gastrointestinal tract. Although cyanide is known to be a rapidly acting poison, its effect is also dose dependent. Acute and subacute toxic effects of cyanide can vary from events like convulsions, screaming, vomiting, coma, and death.[68] The lethal dose of cyanide in adult humans is 50 to 60 mg. Inhalation of HCN at concentrations of 110 to 135 ppm has been reported to produce fatality within a few hours.[69] The actual toxicity depends on the rate of absorption and the ability of the body to dispose of the cyanide.

The effects of long-term exposure to low levels of cyanide are not well understood. Since the beginning of this century, it was known that repeated injections of cyanide could lead to neurological changes in animals.[70] Experimental evidence from feeding trials conducted to study neuropathological effects of cyanide indicated that rats fed with cassava developed neuromuscular symptoms.[71] Lumsden,[72] Rose et al.,[73] Ibrahim et al.,[74] and Smith et al.[75] subjected rats and dogs to sublethal doses of cyanide repeatedly and specific lesions were detected in the central nervous system. Smith and Duckett[76] also obtained an increased level of serum thiocyanate and degeneration of myelin in the central nervous system. Experimentally chronic HCN intoxication was found to lead to irreversible damages to the central nervous system.[77,78] Hurst[79] found that the most apparent lesion in monkeys receiving potassium cyanide (i.m.) was the necrosis of white matter. Experimental cyanide encephalopathy was studied in rats by many workers.[65,80] Lesions were most common in the grey and white matter in these studies. Suzuki[81] reported pathological changes in the contractile components of the myocardium in animals given large doses of cyanide.

2. Action and Metabolism of Linamarin

It is well established that the toxicity of cassava is due to the release of free cyanide from linamarin and lotaustralin. The biological effects of cassava cyanogenic glycosides are however, only poorly documented. In sheep, lotaustralin has been shown to be toxic as the rumen flora can easily hydrolyze it. The lethal dose has been estimated to be 4 to 5 mg/kg body weight.[82] The intestinal hydrolysis of linamarin has been reported by many workers.[62,83] However, Bourdoux et al.[50] found that out of the several microorganisms of the intestines, only *Klebsiella* could effectively hydrolyze linamarin. Fukuba and Mendoza[84] studied the intestinal absorption of linamarin in rats. Everted sections from the middle one third of the intestine were found to absorb linamarin proportionately up to 0.2 M linamarin. Free cyanide was absorbed in huge amounts. These studies also showed that the absorption of linamarin occurred without the involvement of β-glucosidase. Ions like Ca^{2+} and Mg^{2+} were not found to significantly affect the absorption of linamarin and cyanide by the intestine. As compared with amygdalin, 25% more linamarin was absorbed by the rat intestine.[84]

Auld[85] reported no signs of toxicity in guinea pigs administered 150 mg linamarin over a period of 24 hr. Barrett et al.[86] found that linamarin given orally to rats is absorbed and metabolized to yield thiocyanate which is then excreted in the urine. These workers showed that 19% of a single dose of 30 mg/100 g body weight (b.w.) of rat administered orally appeared as the intact glycoside in the first 24 hr, and another 24% was excreted as urinary thiocyanate. The formation of urinary thiocyanate indicates the in vivo hydrolysis of linamarin. An oral dose of 50 mg linamarin was found to be lethal to rats.[87,88] Philbrick et al.[89] also found that several biochemical changes occurred in linamarin-dosed animals. These included severe metabolic acidosis, decreased cytochrome oxidase activity, and decreased respiratory rates. Cardiac arrhythmias indicated by altered electrocardiogram (ECG) patterns were also reported in linamarin-dosed rats.[90] In these rats, decreased cytochrome oxidase

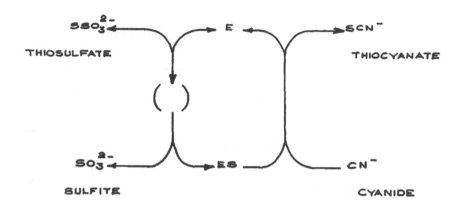

FIGURE 4. Mechanism of rhodanese action.[106]

and ATPase activities were also noted. In a recent study, Ng[91] demonstrated the hydrolysis of several glycosides such as prunasin, amygdalin, dhurrin, taxyphyllin, and linamarin by a β-glucosidase preparation from rabbit liver.

D. In Vivo Detoxification of Cyanide

The ingestion of sublethal doses of cyanide stimulates the defense mechanisms of the body so as to ensure the detoxification of a highly toxic compound to a less toxic product. Several pathways exist for the detoxification of cyanide and these involve several enzymes and cellular components.

The principal detoxification pathway for cyanide is controlled by rhodanese (E.C.2.8.1.1; thiosulfate-cyanide sulfur transferase) which catalyzes the transfer of sulfur from thiosulfate to cyanide forming thiocyanate. Lang[92] was the first to report that injection of cyanide accelerated the excretion of thiocyanate.

$$S_2O_3^{2-} + CN^- \xrightarrow{\text{rhodanese}} SCN^- + SO_3^{2-}$$

This was later confirmed by several workers.[93,94] In 1933, Lang[95] postulated the existence of an enzyme which he called "rhodanese" which was responsible for the formation of thiocyanate from cyanide. Cosby and Sumner[96] purified the enzyme and its kinetic parameters and distribution patterns were studied by others.[97-99] Himwich and Saunders[99] found that the amount of rhodanese in the liver varied with different animals and therefore decided the capacity of various animals to cope with cyanide. According to their studies, the highest levels of rhodanese were present in rats, followed by rhesus monkeys and rabbits, while the lowest levels were found in dogs. The nature of interactions of rhodanese with thiosulfate was further investigated by Saunders and Himwich.[100] Sorbo[101,102] found that rhodanese contained an active disulfide group which participated in the reactions. Rhodanese is widely distributed in the body with the highest concentration in the liver and kidneys.[103] In addition to cyanide detoxification, rhodanese also plays a role in sulfur metabolism.[104,105]

Based on chemical and kinetic data, a detailed mechanism of action has been postulated for rhodanese.[106] Its formal mechanism of action is given in Figure 4.

The thiosulfate for detoxification is provided by sulfur containing amino acids and hence, these are essential for detoxification process.[107]

Meister[108] found that 3-mercaptopyruvic acid arising from the transamination or deamination of cysteine can also act as a sulfur donor for cyanide detoxification. This was confirmed by Wood and Fiedler[109] and a distribution pattern of the enzyme was studied in tissues like liver, kidneys, spleen, pancreas, etc. by Meister et al.[110] (see Figure 5).

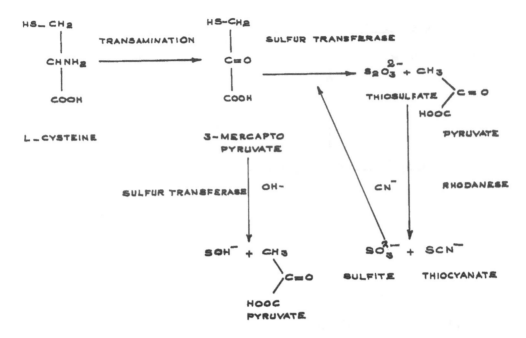

FIGURE 5.   Cyanide-3-mercaptopyruvate cycle.[110]

An alternate pathway for cyanide detoxification is provided by vitamin B_{12} through its ability to form cyanocobalamin from the hydroxocobalamin.[111] In the presence of light, cyanocobalamin is converted to hydroxocobalamin and this further reacts with cyanide to form cyanocobalamin.[112]

Blakeley and Coop[113] reported that cystine and cysteine can also act as sulfur donors for sulfur detoxification through another pathway. Voegthin et al.[114] have found that a dose of cystine injected immediately before the injection of cyanide protected the animals from a minimum lethal dose. Wood and Cooley[115,116] demonstrated the scheme for cyanide detoxification as presented in Figure 6.

Boxer and Rickards[117] have shown that after injection of cyanide labeled with ^{14}C, a fraction of radioactivity is converted to $(^{14}C)\ CO_2$ and 45% of the injected isotope is recovered in the urine after 24 hr.[118]

When hydrogen cyanide is converted to thiocyanate, there is a 200-fold reduction in toxicity.[119] By this efficient mechanism, the body is able to cope with small amounts of cyanide. Goldstein and Rieders[120] reported the presence of a thiocyanate oxidase in the erythrocytes which converted thiocyanate back to cyanide. This may be maintaining an equilibrium between cyanide and thiocyanate in blood.

E. Cyanide and Human Diseases
1. Leber's Optical Atrophy

In humans, two neurological syndromes resulting from chronic exposure to cyanide have been recognized, viz., Leber's optical atrophy and tobacco amblyopia.[121-123] These defects arise due to an impairment in the conversion of cyanide to thiocyanate. Detailed studies on these two syndromes were made by Wilson.[124]

2. Tropical Ataxic Neuropathy

Epidemiological and clinical studies conducted in tropical areas where cassava forms a staple food item have shown that tropical neuropathies characterized by nerve deafness, optical atrophy, and ataxia are common in these areas. The activity of cyanide as an etiologic

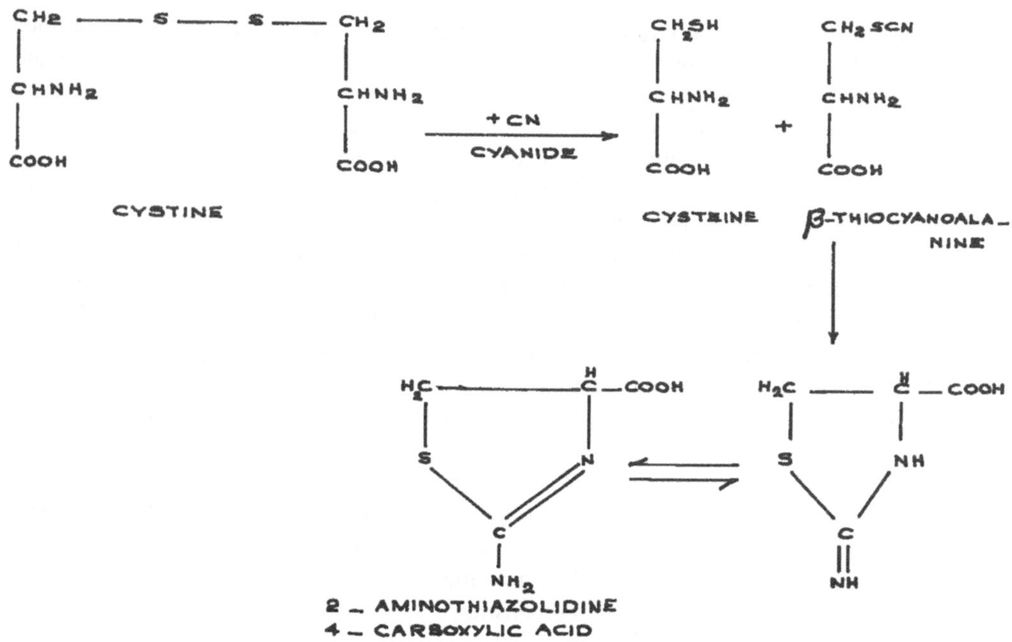

FIGURE 6. Detoxification pathway of cyanide involving cystine.[116]

agent in producing human neuropathies has been investigated in detail by Lessel.[125] Osuntokun et al.[126-128] described the chronic degenerative disease called tropical ataxic neuropathy (TAN) which occurs in several parts of Africa. The disease is prevalent among African communities with a low standard of nutrition and high cassava intake. The syndrome has been henceforth studied by several workers.[129-131] TAN is characterized by myelopathy, bilateral optical atrophy, and perceptive deafness. The symptomatology and epidemiological characters associated with the disease have been reviewed in detail by Osuntokun.[132] There is a close correlation between the amount of cassava consumed and the serum levels of cyanide and thiocyanate. Sulfur containing amino acids were absent in the plasma of 60% of the patients. This deficiency might have arisen from an intense utilization of these in the detoxification of cyanide. Plasma cyanocabalamines, on the other hand, were reported to be elevated. Free cyanide as distinct from total cyanide was present in a higher concentration in the blood of TAN patients compared with normal Nigerians. The activity of rhodanese in the liver of patients was normal. Nerve conduction velocity was reduced in the TAN patients. Chronic cyanide intoxication has been reported to induce demyelination in experimental animals and the primary effect is on glial cells.[133] Osuntokun[134] has found that a cassava diet fed to Wistar rats for 18 to 24 months elevated the plasma thiocyanate levels and ataxic and segmental demyelination were reported in these rats. Dementia has been observed in 9% of TAN patients which might probably have occurred due to callosal demyelination.[127] There was also increased incidence of goiter in Nigerian patients with TAN.[130]

3. Endemic Goiter and Cretinism

Although the goitrogenic properties of plants containing linamarin were known long ago,[135] the goitrogenic effect of cassava was suspected for the first time in 1966 in East Nigeria. The frequency of goiter was so high in these areas that iodine deficiency alone could not account for this high incidence.[136,137] The goitrogenic action of cassava was originally attributed to a thionamide compound.[137] Further studies showed that the antithyroid action of cassava was due to the thiocyanate produced from it.[138-140] The metabolic abnormalities in the thyroid induced in rats fed cassava were similar to those produced by thiocyanate administration.[141] The antithyroid effect of thiocyanate was first demonstrated by Barker[142] in patients with hypertension who were treated with high amounts of thiocyanate. The inhibition by thiocyanate of iodide uptake by the thyroid gland has been clearly demonstrated experimentally.[143-145] Thiocyanate inhibited iodide uptake by increasing the velocity constant of iodide efflux from the gland.[146-148]

Epidemiological studies conducted in Eastern Nigeria[149] clearly indicated the role of cassava in the etiology of endemic goiter. In the goiter patients, there was consistently high serum and urinary thiocyanate.[137,138,149] Clinical studies conducted in areas like Idjwi Island, Kivu, and Ubangi which are all predominantly goiter areas also demonstrated the interrelationship between goiter and cassava intake.[50,150] The results of detailed studies conducted in goiter patients have been summarized in two books by Ermans et al.[151] and Delange et al.[152]

Tewe[153] reported that in growing pigs fed cassava peel rations, there were pathological changes in the colloid and secretory cells of the thyroid gland. There was also increased amniotic fluid thiocyanate in gestating rats fed cassava.[154] Consistently high serum thiocyanate and rhodanese activity occurred in growing rats fed fresh or dry cassava meal.[155] Studies conducted by Ayangade et al.[156] demonstrated increased levels of plasma and amniotic fluid thiocyanate in cassava-eating women.

Prevalence of endemic cretinism is also high in the cassava consuming population.[151] Experimental studies conducted on the mechanisms responsible for mental retardation resulting from cassava ingestion have shown that administration of HCN to pregnant rats in presence of iodine deficiency hindered brain development of their offspring and suckling offsprings.[157] The effect of HCN on cerebral development was shown to be mediated through iodine insufficiency.

4. Tropical Calcifying Pancreatitis

Tropical calcifying pancreatitis (TCP) and pancreatic diabetes (PD) are symptoms often reported from Kerala, India.[158] The disease has also been reported to be prevalent in countries like Nigeria,[128] Uganda,[159] India,[160] Indonesia,[161] and Brazil.[162] Chances of correlation between cassava intake and incidence of the disease have been speculated by these workers. Increased incidence of pancreatic diabetes in cassava-consuming people has been reported.[163] Histopathological studies conducted in rats administered cassava for 18 months[164] also showed pancreatic changes like dilated ductules, papillary infoldings, atrophic acini, round cell infiltration, etc. However, further studies are needed before any definite conclusion can be drawn on the role of cassava in the etiology of TCP and PD.

F. Prophylactic Action of Cassava

The prophylactic action of cassava intake has been reported for the first time by Oke.[165] The metabolites formed from cassava cyanogenic glycosides are physiologically active and can exert some beneficial effects in the body. One of these is the incidence of certain cases of hypertension resulting from low levels of serum thiocyanate. Hydrocyanic acid is required for the formation of cyanocobalamin (vitamin B_{12}). Another disease which can be alleviated partially is the sickle-cell disease. This disease comes in the category of ''molecular diseases'' and can be treated to some extent using cyanate and thiocyanate. The effectiveness of urea in controlling the aggregation of sickle-cell hemoglobin has been proposed to be due to the cyanate contained in it.[166] Manning et al.,[167,168] found that cyanate irreversibly inhibited the sickling of red blood cells in vitro. Under in vivo conditions, the life span of treated sickle-cells remained in the near normal range.[169,170] Traces of cyanide in the foods can be converted to cyanate and thiocyanate within the body. These indirectly can benefit the sickle-cell anemic patients acting as a prophylactic.

One of the most common cancers is that of the large bowel and this has been found to have strong correlation with food habits. The incidence of bowel cancer is very low in African countries.[166,171,172] The neoplastic cells are devoid of the cyanide detoxifying enzyme rhodanese. In turn these cells have a high β-glucosidase activity which can therefore release free cyanide from its bound form at the site of malignancy. The neoplastic cells are then subjected to cytotoxicity by the liberated cyanide which in turn can kill the cancer cells. The somatic cells remain unaffected due to the presence of rhodanese in them. The anticancer property of amygdalin has been well recognized.[173] The anticancer effect of cyanoglycosides has also been refuted by other workers.[65]

Another disease called Schistosomiasis produced by the infection of *Schistosomiasis hematobium* and *S. mansoni* also finds relief through treatment with the cyanoglucoside, amygdalin. Amygdalin was found to reduce the egg production of *S. mansoni* from 100 to 200/day to zero.[166] Although the prophylactic effect of cyanogenic glycosides on certain diseases has been well recognized, this aspect has to be dealt with very carefully, as the contraindications of these glycosides as drugs may sometimes outweigh their beneficial effects.

G. Detoxification of Cassava Through Processing Techniques

Although the toxic properties of cassava are well recognized, cassava is eaten by several million people in the tropics. Traditional processing methods reduce the cyanide content of cassava to an appreciably low level which makes the consumption of cassava relatively safe. Soaking and heat processing are important steps in the processing of cassava tubers.

Traditional detoxification processes adopted with the intention of reducing the hydrocyanic acid (HCN) content vary from country to country. These methods are drying, soaking, boiling, fermentation, or combinations of two or more of these.

1. Drying

Cassava is usually processed into several dried products. The effect of oven drying of chips at 60, 105, and 165°C was investigated by Bourdoux et al.,[174] and the HCN content was found to be reduced at temperatures just above 60°C. The effect of chip thickness on oven drying at 60°C was studied by Gomez et al.[175] The total and bound cyanide were found to be low in thick chips after over drying compared with thin chips. In thick chips, the linamarase activity may be continuing for a longer time as dehydration is at a slower rate, while water which is a substrate for the enzyme is lost at a fast rate from the thin chips. Sundrying on concrete floors has been reported to eliminate more bound cyanide than drying on inclined trays.[175] About one third of the bound cyanide was removed by drying in an air drier at 47 and 60°C. Faster drying at 80 or 100°C eliminated only 10 to 15% of bound cyanide.[176,177] Reports on the extent of cyanide elimination are highly contradicting. Paula and Rangel reported an 85% loss while Charavanapavan reported a 90% loss.[178,179] Tewe and Maner[180] found that 43 and 94% of cyanide was lost from sweet and bitter grated cassava by oven drying. Different heat treatments were found to retain different levels of cyanide (see Table 4).[181]

2. Soaking and Boiling

Soaking 1 cm diced samples for 24 hr followed by 10 min boiling has been reported to eliminate up to 60% of the total cyanide and remove the free cyanide completely.[182] Soaking alone is not effective in removing cyanide, since only 14% reduction in total cyanide was observed.[182] In steaming and microwave cooking, 22 to 78% loss of total cyanide was observed for the former while 44 to 72% reduction was observed for the latter. Large variations in the extent of cyanide elimination exist between varieties.

Among the common cooking practices for cassava leaves, boiling the leaves for a long

Table 4
EFFECT OF DIFFERENT PROCESSING
METHODS ON THE CYANOGLUCOSIDE
RETENTION IN CASSAVA[181]

	CNG content[a] (µg cyanide/g fresh weight)		
Process	Variety H-165	H-2304	H-1687
Fresh	140 ± 4.2	82.5 ± 1.3	58.2 ± 1.2
Boiling	77.6 ± 1.6	43.5 ± 0.8	30.7 ± 0.5
Baking	122 ± 2.8	70.1 ± 1.3	49.6 ± 1.1
Frying	125 ± 2.5	75.2 ± 1.6	49.8 ± 1.4
Steaming	121 ± 2.5	70 ± 1.5	47.5 ± 1.6
Drying	99.2 ± 1.8	60.5 ± 1.4	43.5 ± 0.9

[a]   Values are mean of five samples ± S.E.

time has been reported to eliminate up to 85% total cyanide. Soaking for 1 hr followed by boiling for 10 min has been reported to eliminate 65 to 80% of the cyanide. Up to 82 to 100% reduction in leaf cyanide content could be achieved by crushing and parboiling in water for 15 min, washing, and finally cooking.[183] The studies conducted by Fukuba et al.[182] show that time of heating and sample size are two key factors determining the extent of cyanide elimination. Prolonged soaking of tubers up to 5 days could reduce the cyanide content to 3% of the initial level.[174]

3. Fermentation

In fermented products like gari, fufu, etc., chances of cyanide loss are much higher. During gari preparation, the disintegrated cassava is fermented for 4 days. The extruded juice which is squeezed out after fermentation contains most of the cyanide. Collard and Levi[184] observed that gari fermentation proceeded through two specific stages. *Corynebacterium manihot* initiates fermentation by breaking down cassava starch and producing organic acids which subsequently lower the pH of the mash. At this low pH, the hydrolysis of cyanoglucosides to free HCN occurs. This acid pH then favors the multiplication of *Geotrichum candidum* which produces aldehydes and esters that impart the aroma to gari.[185] According to Ekundayo, the low pH and activity of cassava linamarase will breakdown much of the bound cyanide.[186] Other organisms like *Pseudomonas sp.*, *Bacillus* spp., *Aspergillus* spp., were also isolated from fermenting cassava mash. Cassava linamarase along with microbial β-glucosidases may be acting to detoxify the cyanoglucosides.

The detoxifying role of exogenously added linamarase during gari production was evaluated by Ikediobi and Onyike.[187] They found that enzyme hydrolysis and not low pH was responsible for the detoxification of cyanoglucosides during gari production. The effect of added microorganisms capable of hydrolyzing linamarin during cassava fermentation was investigated by Ejiofor and Okafor.[188] Two organisms *Leuconostoc mesenteroids* and *Alcaligenes faecalis* hydrolyzed linamarin at 72 hr itself, while *Saccharomyces cerevisiae* and *Rhodotorula minuta* hydrolyzed linamarin at 96 hr of growth.[188] Meuser and Smolnik showed that freeze-drying or flash drying eliminated 50% of the total cyanide while roller drying or drum drying of fresh pulp at a pH of 5.5 to 5.8 could eliminate much of the cyanide.[189] Roller drying of fermented pulp resulted in a high residual cyanide. The extent of cyanide elimination also depends on the original cyanide content of fresh roots.

H. Assay Methods for Cyanoglucosides in Cassava and Cassava Products

The presence of cyanogenic glycosides in cassava is of great concern as it is a popular root crop among the tropical countries. Ingestion of cassava can lead to chronic or acute toxicity through liberation of hydrocyanic acid from the cyanoglucosides. Several clinical disorders have been correlated with chronic exposure to sublethal doses of cyanide arising from a prolonged consumption of cassava. These findings necessitate an accurate assay of the cyanoglucoside content of cassava and its products. Cassava leaves are a high protein forage for livestock. Several reports of accidental poisoning of livestock from intake of cassava leaves are available. A critical assay of the cyanoglucoside content of leaves will also help evolve low cyanide varieties of cassava which offer a wide potential as animal feed.

Recent developments in the understanding of cassava linamarase have led to more accurate and sensitive assay methods for the total cyanide content of cassava. Direct assay of cyanogenic glycosides will be too tedious and can lead to unreliable results. Hence, much of the available methods rely on the release of free cyanide from the cyanogenic glycosides either chemically or enzymatically, and then assaying the free cyanide by any common method for determination of cyanide.[190]

Quantitative estimation of cyanide present in a cyanogenic foodstuff necessitates three distinct steps.

1.   Hydrolysis of cyanogenic glycosides to liberate free cyanide.
2.   Isolation and concentration of the liberated HCN.
3.   Determination of cyanide using chemical reaction or physical behavior of the cyanide ion.

1. Liberation of HCN from Cyanogenic Glycosides of Cassava

The preparation of plant material to avoid loss of cyanide (bound or free) is of prime importance in getting an accurate estimation of cyanide.

The primary step in the hydrolytic process is reducing the plant tissue to a fine size. The finer the particles, the better the yield of hydrogen cyanide (HCN).[191] There are three methods for accomplishing hydrolysis. These are (1) acid hydrolysis, (2) enzyme hydrolysis, and (3) combined enzyme and acid hydrolysis. Difficulties often arise in this stage since it is accomplished by either acid hydrolysis or autolysis. Chemical hydrolysis with acids is often subject to errors as the chances of incomplete hydrolysis are high.

Some authors have reported the hydrolysis of cyanogenic glycosides using dilute acids. However, Pulss obtained lower yields of hydrocyanic acid using acid hydrolysis.[192] Hanssen and Sturm also reported lower yields which they found were due to the stabilization of cyanohydrin at low pH values of below 3.5.[193] In the presence of concentrated acids, there is a tendency to form α-hydroxy carbonic acid and ammonium salt.[194] In contrast, a high yield of HCN has been reported by acid hydrolysis by Lehmann and Zinsmeister.[195]

Enzyme hydrolysis is often achieved through the action of endogenous β-glucosidases present in the plant tissue.[191] Autolysis of the plant tissue using endogenous hydrolytic enzymes can lead to erroneous conclusions as the process is entirely dependent on the enzyme levels in the particular tissue, conditions provided for autolysis, long incubation periods required to achieve hydrolysis, etc. During the prolonged incubation period, it is quite likely that secondary reactions using the released cyanide due to the action of endogenous rhodanese, cyanoalanine synthetase, etc. are initiated.[196] In such cases, cyanide normally will escape detection and the results can be erroneous. Autolysis cannot be applied in the case of cassava products where the endogenous linamarase would have been killed by the heating process.

A more accurate enzymic hydrolysis of cyanogenic glucoside is achieved through external β-glucosidase preparations. In using enzyme hydrolysis, some conditions, such as temperature and pH, optimal requirements of the enzyme have to be correctly assessed before proceeding with the hydrolytic step.[64] To avoid long incubation periods, the quantity of plant material is kept as minimum and an excess of enzyme preparation is used.[197]

Cassava linamarase can be prepared from cassava peel either by the acetone precipitation method or by the ammonium sulfate precipitation method.[64,198]

In the acetone precipitation method, the cassava peel is ground to a fine powder and extracted with 0.1 M chilled acetate buffer (pH 5.5) (1:3 w/v). The homogenate is filtered through cheese cloth and centrifuged at 1,500 rpm at 4°C for 15 min. The clear supernatant is used for acetone precipitation. There are 20 vol of the supernatant mixed with 80 vol of prechilled acetone while stirring. The precipitate is collected by decantation and acetone is evaporated by air circulation.

The enzyme homogenate is prepared by dissolving a definite amount of this powder in 0.1 M acetate buffer (pH 5.5).

In ammonium sulfate precipitation method, dried cassava peel is homogenized with 0.1 M acetate buffer (pH 5.5) (1:10 w/v). The homogenate is centrifuged at 10,000 g for 30 min and the supernatant is brought to 60% ammonium sulfate saturation and held at 4°C for 16 hr. The precipitate is collected by centrifugation at 10,000 g for 1 hr and suspended in 0.1 M phosphate buffer (pH 6.0). The suspension is dialyzed against the buffer for 24 hr.

In a combined enzyme and acid hydrolysis method, the enzyme hydrolysis is followed by acid hydrolysis.[199,200] However, Hanssen and Sturm reported lower yield of HCN by this method than with exclusive enzyme hydrolysis.[193]

2. Isolation of HCN

This step can be skipped in the direct assay procedures which offer a more accurate assay of the cyanide content.[84] Two methods are generally used for isolation of the liberated hydrogen cyanide (HCN), viz., aeration and distillation. The HCN isolated by aeration or distillation is trapped in an alkali like sodium hydroxide or sodium carbonate,[201] or directly in alkaline picrate solution.[202] Incomplete trapping of HCN can occur in such experiments. In direct assay methods, the loss of cyanide can be reduced to a minimum and recently a spectrophotometric direct assay is preferred for the determination of cyanide.

3. Spectrophotometric Direct Assay Methods for Cyanide

Spectrophotometric methods for cyanide determination are sensitive, less time consuming, and uncomplicated. An important method is based on Konig reaction, in which cyanogen halides are formed by treatment of cyanide with bromine or chlorine, which are further treated with pyridine to form glutaconic aldehyde.[203] This is then coupled with primary amines or compounds with active methylene groups to form colored compounds. Aldridge[204] used benzidine as amine to form a red color. The carcinogenic nature of benzidine was a handicap of this method and this was later substituted by phenylenediamine. Owing to the strong allergenic nature of this compound, this was later replaced by anthranilic acid. Barbituric acid was used as a coupling agent by Asmus and Garschagen.[205] This procedure was further modified by Lambert et al.[206] and Nambisan and Sundaresan.[207] Another simple method depending on the binding of cyanide to hemoglobin was developed by some other workers.[195] The hemoglobin reagent is highly unstable making the assay method of limited practical value.

The picric acid test most often used for the qualitative identification of cyanide may lead to false results. Ikediobi et al.[198] have modified the alkaline picrate test for the direct assay of cyanide from the reaction medium itself. However, Mendoza et al.[200] have evaluated this technique with other spectrophotometric assays and found that 2 to 10 times greater values are obtained in this method. Their comparative studies of the spectrophotometric methods showed that the pyridine-pyrazolone method[64] and pyridine-barbiturate method[206] gave consistently similar values for cyanide. These workers have attributed the high values of cyanide obtained in the picrate method to the interference of other constituents in plant materials. The stability of the reagents as well as the final colored product is better for the pyridine-barbiturate method than the pyridine-pyrazolone method.[200] Furthermore, considering the availability of chemicals and cost of reagents also, the pyridine-barbiturate method stands superior to the pyridine-pyrazolone method.

An isotacho-electrophoretic method has been used by Mendoza et al.[200] for the determination of cyanide in cassava. The sensitivity limit of the method is 0.1 to 0.36 µg cyanide and can be completed in 30 to 35 min. The cyanide values obtained by this method were similar to those obtained by the pyridine-pyrazolone method. In the isotacho-electrophoretic method, the peeled cassava roots are crushed under pressure and the juice collected and centrifuged. The supernatant is then treated with a few drops of toluene and kept in a refrigerator until use. There are 10 µℓ of the supernatant diluted with water to proper concentration and fed to isotacho-electrophoretic apparatus (Shimadzu IP-IB). Cyanide ion is identified by the potential unit value (PUV). Although the method is fast, accurate, and convenient for handling a number of samples, the need for special instrumentation makes the method of limited practical application.

REFERENCES

1. **Oyenuga, V. A.**, The composition and nutritive value of certain feedingstuffs in Nigeria. I. Roots, tubers and green leaves, *Emp. J. Exp. Agric.*, 23, 81, 1955.
2. **Johnson, R. N. and Raymond, W. D.**, The chemical composition of some tropical food plants. IV. Manioc, *Trop. Sci.*, 7, 109, 1965.
3. **Oke, O. L.**, Cassava as food in Nigeria, *World Rev. Nutr. Diet*, 9, 272, 1968.
4. **Mesa, J., Maner, J. H., Opando, H., Portela, R., and Callo, J. T.**, Nutritive value of different tropical sources of energy, *J. Anim. Sci.*, 31, 208, 1970.
5. **Hutagalung, R. I.**, Nutritive value of leaf meal, tapioca root meal, normal maize and opaque-2, maize and pineapple bran for pig and poultry, 17th Annu. Conf. Mal. Vet. Assoc., University of Malaya, Kuala Lunpur Malaysia, 1972.
6. **Muller, Z., Chou, K. C., Nah, K. C., and Tan, T. K.**, Study of nutritive value of tapioca in economic rations for growing finishing pigs in the tropics in United Nations Developmental Program (UNDP), SF Project SIN 67/505, 672, 1, 1972.
7. **Omole, T. A.**, Cassava in the nutrition of Layers, in *Cassava as Animal Feed, Proc. Workshop*, Nestel, B. and Graham, M., Eds., University of Guelph, Ontario, Canada, 1977, 51.
8. **Hutagalung, R. I., Phuah, C. H., and Hew, V. F.**, The utilization of cassava, tapioca (*Manihot utilissima*) in livestock feeding, in *Proc. Third Symp. Trop. Root Tuber Crops*, 1973, 45.
9. **Muller, Z., Chou, K. C., and Nah, K. C.**, Cassava as a total substitute for cereals in livestock and poultry rations, in *Proc. 1974 Trop. Products Institute Conf.*, 1975, 85.
10. **Vogt, H.**, The use of tapioca meal in poultry rations, *World's Poult. Sci. J.*, 22, 113, 1966.
11. **Ketiku, A. O. and Oyemuga, V. A.**, Preliminary report on the carbohydrate constituents of cassava root and yam tuber, *Niger, J. Sci.*, 4, 25, 1970.
12. **Sreeramamurthy, V. V.**, Investigations on the nutritive value of tapioca, *Ind. J. Med. Res.*, 33, 229, 1945.
13. **Montaldo, A.**, Whole plant utilization of cassava for animal feed, in *Cassava as Animal Feed, Proc. Workshop*, Nestel, B. and Graham, M., Eds., University of Guelph, Ontario, Canada, IDRC-095e, 1977, 95.
14. **Maini, S. B.**, Quality aspects of cassava, in *Cassava Production Technology*, Hrishi, N. and Nair, R. G., Eds., Central Tuber Crops Research Institute, Trivandrum, India, 1978, 49.
15. **Maner, J. H.**, Cassava in swine feeding, in Centro International de Agricultura Tropical, Cali, Colombia, Bulletin RB-1, 1973.
16. **Khajarern, S. and Khajarern, J. M.**, Use of cassava as a food supplement for broiler chicks, in *Proc. Fourth Symp. Inter. Soc. Trop. Root Crops*, Cock, J., MacIntyre, R., and Graham, M., Eds., CIAT, Colombia, IDRC-080 e, 1977, 246.

17. **Hudson, B. J. F. and Ogunsua, A. O.,** Lipids of cassava tubers (*Manihot esculenta* Crantz), *J. Sci. Food Agric.,* 25, 1503, 1974.

18. **Barrios, E. A. and Bressani, R.,** Composicion guimica de la raiz y de la hojas de algunas variedades de yuca Manihot, *Turrialba,* 17, 314, 1967.

19. **Rojas, I.,** Vitamina C en yucas peruanas y su variacion por coccion, Lima, Peru, Univ. Nac. Sn. Marcos, 1968, 56.

20. **Obregon, R. and Juarez, G.,** Comparacion entra follaje de yuca y alfalfa, *Estac. Exp.,* 1955.

21. **Bangham, W. N.,** A mandioca supera a alfalfa, *Fezenda,* 45, 27, 1950.

22. **Echandi, M. O.,** Value of dehydrated cassava leaf and stalk meal for milk production, *Turrialba,* 2, 166, 1952.

23. **Juarez, G. L.,** The utilization of cassava leaves as forage, Lima, Estacion Experimental Agricola de La Molina Bol., Lima, Peru, 1955, 58.

24. **Miranda, R. M., De, Laun, G. F., and Costa, B. I. Da.,** Use of Cassava Meal, Tropical Kudzu, *Desmodium dsicolor* and Alfalfa Hay in Chick Rations, *Instituto de Zootecnis Publicacao* No. 19, 1957, 18.

25. **Sen, K. C., Ray, S. N., and Ranjhan, S. K.,** *Nutritive Value of Indian Cattle Feeds and the Feeding of Animals,* Indian Council of Agricultural Research, New Delhi, 1978.

26. **Paula, R. D. de, G. and Cavalcanti,** Bol. Inst. Nac. Tech. (Riode J), 3, 21, 1952.

27. **Gramacho, D. D.,** Chemical study of cassava as forage, *Ser. Pesquisa,* 1, 143, 1973..

28. **Pechnik, E., Guimaraes, L. R., and Panek, A.,** Sobre o aproveitamentao da folha de manioca, (*Manihot* sp.) na alimentaciao humana. II. Contribucao ao estudo do valor alimenticio, *Arg. Bras. Nutr.,* 18, 11, 1962.

29. **Luyken, R., Groot, A. P., De., and Van Stratum, P. G. C.,** Nutritional value of foods from New Guinea. II. Net Protein utilization, digestibility and biological value of sweet potatoes, sweet potato leaves and cassava leaves from New Guinea, Utrecht, 1961, 18.

30. *Production Yearbook,* Vol. 29, Food and Agriculture Organization, Rome, 1975.

31. **Farelo, De R.,** Meal of cassava leaves and stalks in animal feeding, *Chacaras Quintais,* 114, 663, 1966.

32. **Conceicao, A. J., Da, Sampaio, C. V., and Mendes, M. A.,** Competicao de variedades de aipim e mandioca para forragem, University Federal, Bahia (Projecto Mandioca), 1973, 130.

33. **Scholz, H. K. B. W.,** Possibilidades de aproveitamento de partes acreas da mandioca como forrageira, Fortaleza, Banco del Nordeste do Brasil, 1972, 181.

34. **Meyreles, L. and McLeod, N. A.,** La yuca como fuente de proteina forrajera, efecto de la poblacion de plantas y de la edad del corte, Reunion Annual, Santo Domingo, 1976, 2.

35. **Montaldo, A. and Montilla, J. J.,** Produccion de follaje de yuca, *Rev. Fac. Agron.,* 24, 35, 1976.

36. **Clusius, C.,** Exoticorum: libridecem, 1605.

37. **Henry, O. and Boutron-Charland, L.,** Recherches sur le principle veneneux du manioc amer, *Mem. Acad. Med. Paris,* 5, 212, 1836.

38. **Francis, E.,** On prussic acid from cassava, *Analyst,* 2, 4, 1878.

39. **Carmody, A.,** Prussic acid in sweet cassava, *Lancet,* ii, 736, 1900.

40. **Collens, A. E.,** Bitter and sweet cassava — hydrocyanic acid contents, *Bull. Agric. Trim. Tob.,* 14, 54, 1915.

41. **Turnock, B. J. W.,** An investigation of the poisonous constituents of sweet cassava (*Manihot utilissima*) and the occurrence of hydrocyanic acid in foods prepared from cassava, *J. Trop. Med. Hyd.,* 40, 65, 1937.

42. **Cerighelli, R.,** Cultures tropicales, I. Plantes Vivrieres. Bailliere et Fils, Paris, 1955.

43. **Dunstan, W. R. and Henry, T. A.,** Cyanogenesis in plants. V. The occurrence of phaseolunatin in Cassava (*Manihot aipi and Manihot utilissima*), *Proc. R. Soc. London Ser. B,* 78, 152, 1906.

44. **Jorrissen, A. and Hairs, E.,** La linamarine. Nouveau glucoside fournissant de l'acide cyanhydrique par redoublement et retire du *Linum usitissimum, Bull. Cl. Sci. Acad. R. Sci. Belg.,* 21, 529, 189.

45. **Dunstan, W. R. and Henry, T. A.,** Cyanogenesis in plants. III. On phaseolunatin, the cyanogenetic glucoside of *Phaseolus lunatus, Proc. R. Soc. London, Ser. B,* 72, 285, 1903.

46. **Bulter, G. W.,** The distribution of the cyanoglucosides linamarin and lotaustralin in higher plants, *Phytochemistry,* 4, 127, 1965.

47. **Nartey, F.,** Studies on cassava, *Manihot utilissima* Pohl. I. Cyanogenesis: the biosynthesis of linamarin and lotaustralin in etiolated seedlings, *Phytochemistry,* 7, 1307, 1968.

48. **Bissett, F. H., Clapp, R. C., Coburn, R. A., Ettlinger, M. G., and Long, L., Jr.,** Cyanogenesis in manioc: concerning lotaustralin, *Phytochemistry,* 8, 2235, 1969.

49. **De Bruijn, G. H.,** The cyanogenic character of cassava (*Manihot esculenta*), in *Chronic Cassava Toxicity,* Proc. Interdisciplinary Workshop, Nestel, B. and Mac Intyre, R., Eds., London, IDRC, Ottawa, IDRC-010 e, 1973, 43.

50. **Bourdoux, P., Mafuta, M., Hanson, A., and Ermans, A. M.,** Cassava toxicity: the role of linamarin, in *Role of Cassava in the Etiology of Endemic Goitre and Cretinism,* Ermna, A. M., Mbulamoko, N. M., Delange, F., and Ahluwalia, R., Eds., IDRC-136 e, Ottawa, 1980, chap. 1.

51. **Bolhuis, G. G.,** The toxicity of cassava roots, *Neth. J. Agric. Sci.*, 2, 176, 1954.

52. **Periera, A. A. and Pinto, M. G.,** Determinacao du toxicidade de manioca pelo potadar das raices in natura, *Bragantia,* 21, 145, 1962.

53. **Koch, L.,** Cassava Eselectie, thesis, Venman and Zomen, Wageningen, Holland, 1933.

54. **De Bruijn, G. H.,** Etude du caractere cyanogenetique du manioc (*Manihot esculenta* Crantz.). *Meded. Landbouwhogesch. Wageningen,* Holland, 71, 1, 1971.

55. **Bediako, M. K. B., Tapper, B. A. and Pritchard, G. G.,** Metabolism synthetic site and translocation of cyanogenic glycosides in cassava, in *Proc. Triennial Symp. Intern. Soc. Trop. Root Crops,* Terry, E. R., Odure, K. A., and Caveness, F., Eds., IDRC, Ottawa, IDRC-163 e, 1981, 143.

56. **Finnemore, H. and Gledhill, W. C.,** The presence of cyanogenic glucosides in certain species of acacia, *Aust. J. Pharm.*, 9, 174, 1928.

57. **Cooper-Driver, G. A. and Swain, T.,** Cyanogenic polymorphism in bracken in relation to herbivore predation, *Nature (London),* 260, 604, 1976.

58. **Seely, M. K., Criddle, R. S., and Conn, E. E.,** On the requirement of hydroxynitrile lyase for flavin, *J. Biol. Chem.*, 241, 4457, 1966.

59. **Kojima, M., Poulton, J. E., Thayer, S. S., and Conn, E. E.,** Tissue distributions of dhurrin and of enzymes involved in its metabolism in leaves of *Sorghum bicolor, Plant Physiol.*, 63, 1022, 1979.

60. **Conn, E. E.,** Cyanogenic glycosides, *Agric. Food Chem.*, 17, 519, 1969.

61. **Hosel, W. and Conn, E. E.,** The aglycone specificity of plant β-glucosidases, *Trends Biochem. Sci.,* 7, 219, 1982.

62. **Butler, G. W., Bailey, R. W., and Kennedy, L. D.,** Studies on the glucosidase "Linamarase", *Phytochemistry,* 4, 369, 1965.

63. **Wood, T.,** The isolation, properties and enzymic breakdown of linamarin from cassava, *J. Sci. Food Agric.,* 17, 85, 1966.

64. **Cooke, R. D., Blake, G. G., and Battershill, J. M.,** Purification of cassava linamarase, *Phytochemistry,* 17, 381, 1978.

65. **Solomonson, L. P.,** Cyanide as a metabolic inhibitor, in *Cyanide in Biology,* Vennesland, B., Conn, E. E., Knowles, C. J., Westley, J., and Wissing, F., Eds., Academic Press, New York, 1981, 11.

66. **Isom, G. E., Liu, D. H. W., and Way, J. L.,** Effect of sublethal doses of cyanide on glucose metabolism, *Biochem. Pharmacol.*, 24, 871, 1975.

67. **Albaum, H. G., Tepperman, J., and Bodanzky, O.,** The *in vivo* inactivation by cyanide of brain cytochrome oxidase and its effect on the high energy phosphorus compounds of brain, *J. Biol. Chem.,* 164, 45, 1946.

68. **Halstrom, F. and Moller, K. D.,** Content of cyanide in human organs from cases of poisoning with cyanide taken by mouth, with contribution to toxicology of cyanides, *Acta Pharmacol. Toxicol.*, 1, 18, 1945.

69. **Fassett, D. W.,** in *Industrial Hygiene and Toxicology,* Vol. 2, 2nd ed., Fassett, D. W. and Irish, D. D., Eds., John Wiley & Sons, New York, 1963, 191.

70. **Swysen, M.,** Analyse bibliographique sur les thiocyanates. Memoire de Licence on Medecine du Travail, ULB, 1978.

71. **Martino, G.,** On the nutritive value of cassava, *Bull. Inst. Fisiol.*, Asuncion (Paraguay) No. 15, 1935.

72. **Lumsden, C. E.,** Cyanide leucoencephalopathy in rats and observations on the vascular and ferment hypothesis of demyelinating diseases, *J. Neurol. Neurosurg. Psychiatry,* 13, 1, 1950.

73. **Rose, C. L., Harris, P. N., and Chen, K. K.,** Effect of cyanide poisoning on the central nervous system of rats and dogs, *Proc. Soc. Exp. Biol. Med.*, 87, 632, 1954.

74. **Ibrahim, M. L. M., Briscore, P. B., Jr., Bayliss, O. B., and Adams, C. W. M.,** The relationship between the enzyme activity and neuralgia in the prodronial and demyelinating stages of cyanide encephalopathy in the rat, *J. Neurol. Neurosurg. Psychiatry,* 26, 479, 1963.

75. **Smith, A. D., Duckett, S., and Waters, A. H.,** Neuropathological changes in chronic cyanide intoxication, *Nature (London),* 200, 179, 1963.

76. **Smith, A. D. and Duckett, S.,** Cyanide, Vitamin B_{12} experimental demyelination, and tobacco amblyopia, *Br. J. Exp. Pathol.*, 46, 615, 1965.

77. **Hurst, E. W.,** Experimental demyelination of the central nervous system. I. The excephalopathy produced by potassium cyanide, *Aust. J. Exp. Biol. Med. Sci.*, 18, 201, 1940.

78. **Brierley, J. B., Brown, A. W., and Calverley, J.,** Cyanide intoxication in the rat: physiological and neuropathological aspects, *J. Neurol. Neurosurg. Psychiatry,* 39, 129, 1976.

79. **Hurst, E. W.,** Experimental demyelination of the central nervous system. III. Poisoning with potassium cyanide, sodium azide, hydroxylamine narcotics, carbon monoxide, etc. with some consideration of bilateral necrosis occurring in the basal nuclei, *Aust. J. Exp. Biol. Med.*, 20, 297, 1942.

80. **Hicks, S. P.,** Brain metabolism *in vivo.* I. The distribution of lesions caused by cyanide poisoning, insulin hypoglycaemia, asphyxia in nitrogen and fluoracetate poisoning in rats, *Arch. Pathol.*, 49, 111, 1950.

81. **Suzuki, T.** (Quoted by Way, J. L.), Pharmacologic aspects of cyanide and its antagonism, in *Cyanide in Biology*, Vennesland, B., Conn, E. E., Knowles, C. J., Westley, J., and Wessing, F., Eds., Academic Press, New York, 1981, 29.

82. **Coop, I. E. and Blakley, R. L.,** The metabolism and toxicity of cyanides and cyanogenetic glucosides in sheep. I. Activity in the rumen, *N.Z. J. Sci. Technol.*, 30, 277, 1949.

83. **Winkler, W. O.,** Report on methods for glucosidal HCN in limabeans, *J. Assoc. Agric. Chem.*, 41, 282, 1958.

84. **Fukuba, H. and Mendoza, E. T.,** Intestinal absorption of linamarin — the main cyanogenic glucoside in cassava, in *Tropical Root Crops — Postharvest Physiology and Processing*, Uritani, I. and Reyes, E. D., Eds., Japan Scientific Societies Press, Tokyo, 1984, 313.

85. **Auld, S. J. M.,** The formation of prussic acid from linseed cake and other feeding stuffs, *J. Board Agric.*, 19, 446, 1912.

86. **Barrett, M. D. P., Hill, D. C., Alexander, J. C., and Zitnak, A.,** Fate of orally dosed linamarin in the rat, *Can. J. Physiol. Pharmacol.*, 55, 134, 1977.

87. **Barrett, M. D. P.,** Dietary Cyanide, Linamarin and Nutritional Deficiencies, Ph.D. thesis, University of Guelph, Ontario, Canada, 1976.

88. **Barrett, M. D. P., Alexander, J. C., and Hill, D. C.,** Utilization of ^{35}S from radioactive methionine of sulphate in the detoxification of cyanide by rats, *Nutr. Metab.*, 22, 51, 1978.

89. **Philbrick, D. J., Hill, D. C., and Alexander, J. C.,** Physiological and biochemical changes associated with linamarin administration to rats, *Toxicol. Appl. Pharmacol.*, 42, 539, 1977.

90. **Hill, D. C.,** Physiological and biochemical responses of rats given potassium cyanide or linamarin, in *Cassava as Animal Feed, Proc. Workshop*, Nestel, B. and Graham, M., Eds., IDRC-095 e, 1977, 33.

91. **Ng, S. C. J.,** MS thesis, University of California, Davis, 1975.

92. **Lang, S.,** Die Rhodanbildnny in Tierkorper, *Biochem. Z.*, 259, 243, 1933.

93. **Heymanns, J. F. and Mesoin, P.,** (Quoted by Chen, K. K., et al.), Comparative values of several antidotes in cyanide poisoning, *J. Pharmacol. Exp. Ther.*, 51, 761, 1934.

94. **Kahn, M.,** Biochemical Studies of Sulfocyanates, Ph.D. thesis, Columbia University, Indiana, 1912.

95. **Lang, S.,** Die Rhodanbildnny in Tierkorper, *Biochem. Z.*, 259, 262, 1933.

96. **Cosby, E. L. and Sumner, J. B.,** Rhodanese, *Arch. Biochem.*, 7, 457, 1945.

97. **Mendel, B., Rudney, H., and Bowman, M. C.,** Effect of salts on true cholinesterase, *Cancer Res.*, 6, 495, 1946.

98. **Bernard, H., Gadjos-Torok, M., and Gadjos, A.,** Surnn principe necessaire a loction acticyanure de hyposulfite de sodium, *C. R. Soc. Biol.*, 141, 700, 1947.

99. **Himwich, W. A. and Saunders, J. P.,** Enzymatic conversion of cyanide to thiocyanate, *Am. J. Physiol.*, 153, 348, 1948.

100. **Saunders, J. P. and Himwich, W. A.,** Properties of the trans sulfurase responsible for the conversion of cyanide to thiocyanate, *Am. J. Physiol.*, 163, 404, 1950.

101. **Sorbo, B. H.,** Crystalline rhodanese. II. The enzyme catalysed reaction, *Acta Chem. Scand.*, 5, 1218, 1951.

102. **Sorbo, B. H.,** Studies on rhodanese, *Acta Chem. Scand.*, 7, 238, 1953.

103. **Auriga, M. and Koj, A.,** Protective effect of rhodanese on the respiration of isolated mitochondria intoxicated with cyanide. *Bull. Acad. Pol. Sci. (Biol.)*, 5, 305, 1975.

104. **Westley, J.,** Rhodanese, *Adv. Enzymol. Relat. Areas Mol. Biol.*, 39, 327, 1973.

105. **Volini, M. and Alexander, K.,** Multiple forms and multiple reactions of the rhodaneses, in *Cyanide in Biology*, Vennesland, B., Conn, E. E., Knowles, C. J., Westley, J., and Wissing, F., Eds., Academic Press, New York, 1981, 77.

106. **Schlesinger, P. and Westley, J.,** An expanded mechanism for rhodanese catalysis, *J. Biol. Chem.*, 249, 780, 1974.

107. **Wheeler, J. L., Hedges, D. A., and Till, A. R.,** A possible effect of cyanogenic glucoside in sorghum on animal requirements for sulfur, *J. Agric. Sci.*, 84, 377, 1975.

108. **Meister, A.,** Preparation and enzymatic reactions of the Ketoanalozines of asparagine and glutamine, *Fed. Proc.*, 12, 245, 1953.

109. **Wood, J. L. and Fiedler, H.,** β-mercapto pyruvate, a substrate for rhodanese, *J. Biol. Chem.*, 205, 231, 1953.

110. **Meister, A., Fraser, P. E., and Tice, S. V.,** Enzymatic desulfuration of β-mercaptopyruvate to pyruvate, *J. Biol. Chem.*, 206, 561, 1954.

111. **Wokes, F., Baxter, N., Horsford, J., and Preston, B.,** Effect of light on vitamin B_{12}, *Biochem. J.*, 53, 19, 1951.

112. **Kaczka, E. A., Denkewalter, R. G., and Folker, A.,** Vitamine B_{12}. XIII. Additional data on vitamin B_{12}a, *J. Am. Chem. Soc.*, 73, 335, 1950.

113. **Blakeley, R. L. and Coop, I. E.,** The metabolism and toxicity of cyanides and glycosides in sheep. II. Detoxification of hydrocyanic acid, *N.Z. J. Sci. Technol.*, 31, 1, 1949.

114. **Voegthin, O., Johnson, J. M., and Dyer, H. A.**, Glutathione content of turnover animals, *J. Pharmacol. Exp. Ther.*, 27, 467, 1926.

115. **Wood, J. L. and Cooley, S. L.**, Thiocyanate formation from cysteine derivatives, *Fed. Proc.*, 11, 314, 1952.

116. **Wood, J. L. and Cooley, S. L.**, Detoxification of cyanide by cystine, *J. Biol. Chem.*, 218, 449, 1956.

117. **Boxer, G. E. and Rickards, J. C.**, Studies on the metabolism of the carbon of cyanide and thiocyanate, *Arch. Biochem. Biophys.*, 39, 7, 1952.

118. **Crawley, F. E. H. and Goddard, E. A.**, Internal dose from carbon-14 labelled compounds. The metabolism of carbon-14 labelled potassium cyanide in the rat, *Health Physiol.*, 32, 135, 1977.

119. **Oke, O. L.**, The mode of cyanide detoxication, in *Chronic Cassava Toxicity, Proc. Interdisciplinary Workshop*, Nestel, B. and Mac Intyre, R., Eds., London, IDRC., Ottawa, IDRC-010 e, 1973, 97.

120. **Goldstein, F. and Rieders, F.**, Conversion of thiocyanate to cyanide by erythrocytic enzyme, *Am. J. Physiol.*, 173, 287, 1953.

121. **Monekosso, G. L. and Wilson, J.**, Plasma thiocyanate and vitamin B_{12} in Nigerian patients with degenerative neurological disease, *Lancet*, 1, 1062, 1966.

122. **Wokes, F.**, Tobacco amblyopia, *Lancet*, 2, 526, 1958.

123. **Wilson, J., Linnell, J. C., and Matthews, D. M.**, Plasmacobalamines in neuro-ophthalmological disease, *Lancet*, 1, 259, 1971.

124. **Wilson, J.**, Leber's hereditary optic atrophy: a possible defect of cyanide metabolism, *Clin. Sci.*, 29, 505, 1965.

125. **Lessell, S.**, Experimental cyanide optic neuropathy, *Arch. Ophthalmol.*, 84, 194, 1971.

126. **Osuntokun, B. O., Durowoju, J. F., McFarlane, H., and Wilson, J.**, Plasma aminoacids in the Nigerian nutritional ataxic neuropathy, *Br. Med. J.*, 3, 647, 1968.

127. **Osuntokun, B. O., Monekosso, G. L., and Wilson, J.**, Relationship of a degenerative neuropathy to diet. Report of a field survey, *Br. Med. J.*, 1, 547, 1969.

128. **Osuntokun, B. O., Aladetoyinbo, A., and Adeiya, A. O. G.**, Free cyanide levels in tropical ataxic neuropathy, *Lancet*, 2, 372, 1970.

129. **Osuntokun, B. O. and Monekosso, G. L.**, Degenerative tropical neuropathy and diet, *Br. Med. J.*, 3, 178, 1969.

130. **Osuntokun, B. O.**, Epidemiology of tropical nutritional neuropathy in Nigerians, *Trans. R. Soc. Trop. Med. Hyg.*, 65, 454, 1971.

131. **Makene, W. J. and Wilson, J.**, Biochemical studies in Tanzanian patients with tropical ataxic neuropathy, *J. Neurol. Neurosurg. Psychiatr.*, 35, 31, 1972.

132. **Osuntokun, B. O.**, Ataxic neuropathy associated with high cassava diets in West Africa, in *Chronic Cassava Toxicity, Proc. Interdisciplinary Workshop*, Nestel, B. and MacIntyre, R., Eds., London, IDRC, Ottawa, IDRC-010 e, 1973, 127.

133. **Bass, N. H.**, Pathogenesis of myelin lesions in experimental encephalopathy. A microchemical study, *Neurology*, 18, 167, 1968.

134. **Osuntokun, B. O.**, Cassava diet and cyanide metabolism in Wistar rats, *Br. J. Nutr.*, 24, 377, 1970.

135. **Care, A. D.**, Goitrogenic activity in linseed, *N.Z. J. Sci. Technol.*, 36, 321, 1954.

136. **Ekpechi, O. L., Dimitriadou, A., and Fraser, R.**, Goitrogenic activity of cassava (a staple Nigerian food), *Nature (London)*, 210, 1137, 1966.

137. **Ekpechi, O. L.**, Pathogenesis of endemic goitre in Eastern Nigeria, *Br. J. Nutr.*, 21, 537, 1967.

138. **Delange, F. and Ermans, A. M.**, Role of a dietary goitrogen in the etiology of endemic goitre on Idjwi Island, *Am. J. Clin. Nutr.*, 24, 1354, 1971.

139. **Delange, F., Van der Velden, M., and Ermans, A. M.**, Evidence of an antithyroid action of cassava in man and in animals, in *Chronic Cassava Toxicity, Proc. Interdisciplinary Workshop*, Nestel, B. and Mac Intyre, R., Eds., London, IDRC, Ottawa, IDRC-010 e, 1973, 147.

140. **Ermans, A. M., Van der Velden, M., Kinthaert, J., and Delange, F.**, Mechanism of the goitrogenic action of cassava, in *Chronic Cassava Toxicity, Proc. Interdisciplinary Workshop*, Nestel, B. and Mac Intyre, R., Eds., London, IDRC, Ottawa, IDRC-010 e, 1973, 153.

141. **Van der Velden, M., Kinthaert, J., Orts, S., and Ermans, A. M.**, A preliminary study on the action of cassava on thyroid iodine metabolism in rats, *Br. J. Nutr.*, 30, 511, 1973.

142. **Barker, M. H.**, The blood cyanates in the treatment of hypertension, *J.A.M.A.*, 106, 762, 1936.

143. **Vanderlaan, W. P. and Bissell, A.**, Effects of propylthiouracil and of potassium thiocyanate on the uptake of iodine by the thyroid gland of the rat, *Endocrinology*, 39, 157, 1946.

144. **Wolff, J., Chaikoff, I. L., Taurog, A., and Rubin, L.**, The disturbance in iodine metabolism produced by thiocyanate. The mechanism of its goitrogenic action with radioactive iodine as indicator, *Endocrinology*, 39, 140, 1946.

145. **Wollman, S. H.**, Inhibition by thiocyanate of accumulation of radioiodine by thyroid gland, *Am. J. Physiol.*, 203, 517, 1962.

146. **Halmi, N. S.**, Thyroidal iodide transport, *Vitam. Horm.*, 19, 133, 1961.

147. **Halmi, N. S., King, L. T., Winder, R. R., Hass, A. C., and Stuelke, R. J.,** Renal excretion of radioiodide in rats, *Am. J. Physiol.,* 193, 379, 1958.

148. **Scranton, J. R., Nissen, W. M., and Halmi, N. S.,** The kinetics of the inhibition of thyroidal iodide accumulation by thiocyanate: a reexamination, *Endocrinology,* 85, 603, 1969.

149. **Nwokolo, C., Ekpechi, O. L., and Nwokolo, U.,** New foci of endemic goitre in Eastern Nigeria, *Trans. R. Soc. Trop. Med. Hyg.,* 60, 97, 1966.

150. **Delange, F., Vigneri, R., Trimarchi, F., Filetti, S., Pezzino, V., Squatrito, S., Bourdoux, P., and Ermans, A. M.,** Etiological factors of endemic goitre in north-eastern Sicily, *J. Endocrinol. Invest.,* 2, 137, 1978.

151. **Ermans, A. M., Mbulamoko, N. M., Delange, F., and Ahluwalia, R.,** *Role of Cassava in the Etiology of Endemic Goitre and Cretinism,* IDRC, Ottawa, IDRC-136 e, 1980.

152. **Delange, F., Iteke, F. B., and Ermans, A. M.,** *Nutritional Factors Involved in the Goitrogenic Action of Cassava,* IDRC, Ottawa, IDRC-184 e, 1982.

153. **Tewe, O. O.,** Thyroid cassava toxicity in animals, in *Cassava Toxicity and Thyroid: Research and Public Health Issues, Proc. Workshop,* Delange, F. and Ahluwalia, R., Eds., IDRC, Ottawa, IDRC-207 e, 1983, 114.

154. **Tewe, O. O., Maner, J. H., and Gomez, G.,** Influence of cassava diets on placental thiocyanate transfer, tissue rhodanese activity and performance of rats during gestation, *J. Sci. Food Agric.,* 28, 750, 1977.

155. **Tewe, O. O. and Maner, J. H.,** Cyanide, protein and iodide interaction in the performance, metabolism and pathology of pigs, *Res. Vet. Sci.,* 29, 271, 1980.

156. **Ayangade, S. O., Oyelola, O. O., and Oke, O. L.,** A preliminary study of amniotic and serum thiosulphate levels in cassava eating women, *Nutr. Rep. Int.,* 26, 73, 1982.

157. **Colinet, E., Ermans, A. M., and Delange, F.,** Experimental study of mechanisms responsible for mental retardation resulting from cassava ingestion, in *Nutritional Factors Involved in the Goitrogenic Action of Cassava,* Delange, F., Iteke, F. B., and Ermans, A. M., Eds., IDRC, Ottawa, 1982, chap. 9.

158. **Sarles, H. and Laugier, R.,** Alcoholic pancreatitis, *Clin. Gastroenterol.,* 10, 401, 1981.

159. **Shaper, A. G.,** Aetiology of chronic pancreatic fibrosis with calcification seen in Uganda, *Br. Med. J.,* 2, 1607, 1964.

160. **Geevarghese, P. J.,** Cassava diet, tropical calcifying pancreatitis and pancreatic diabetes, in *Cassava Toxicity and Thyroid: Research and Public Health Issues, Proc. Workshop,* Delange, F. and Ahluwalia, R., Eds., IDRC, Ottawa, 1983, 77.

161. **Zuidema, P. J.,** Cirrhosis and disseminated calcification of the pancreas in patients with malnutrition, *Trop. Geogr. Med.,* 11, 70, 1959.

162. **Dani, R. and Nogueira, C. E. D.,** Chronische kalzifizierende Pankreatitis in Braselien, eine Analyse von 92 Fallen, *Leber Magen Darm,* 6, 272, 1976.

163. **Davidson, J. C., Miglashan, M. B., Nightingale, E. A., and Upadhyay, J. M.,** The prevalence of *diabetes mellitus* in the kalene hill area of Zambia, *Med. Proc.,* 15, 426, 1969.

164. **Pushpa, M.,** Chronic Cassava Toxicity — an Experimental Study, M.D. thesis, University of Kerala, India, 1980.

165. **Oke, O. L.,** The role of hydrocyanic acid in nutrition, *World Rev. Nutr. Diet.,* 11, 170, 1969.

166. **Oke, O. L.,** The prophylactic action of cassava, in *Proc. Fourth Symp. Intern. Soc. Trop. Root Crops,* Cock, J., Mac Intyre, R., and Graham, M., Eds., IDRC-080 e, Ottawa, 1976, 232.

167. **Cerami, A. and Manning, J. M.,** Potassium cyanide as an inhibitor of the sickling of erythrocytes *in vitro, Proc. Natl. Acad. Sci.,* 68, 1180, 1971.

168. **Manning, J. M., Cerami, A., Gillette, P. N., de Furia, F. G., and Miller, D. R.,** Chemical and biological aspects of the inhibition of red blood sickling by cyanate, *Adv. Exp. Med. Biol.,* 28, 253, 1972.

169. **Gillette, P. N., Manning, J. M., and Cerami, A.,** Increased survival of sickle-cell erythrocytes after treatment *in vitro* with sodium cyanate, *Proc. Natl. Acad. Sci.,* 68, 2791, 1971.

170. **Cerami, A.,** Cyanate as an inhibitor of red cell sickling, *N. Engl. J. Med.,* 287, 807, 1972.

171. **Krebs, E. T., Jr. and Mc Naughton, A. R. L.,** *Control of Cancer,* Paper Back Library, New York, 1963.

172. **Burkitt, D. P.,** Epidemiology of cancer of the colon and rectum, *Cancer,* 28, 3, 1971.

173. **Krebs, E. T., Jr.,** The nitrilosides (Vitamin B 17), their nature, occurrence and metabolic significance, *J. Appl. Nutr.,* 22, 75, 1970.

174. **Bourdoux, P., Seghers, P., Mafuta, M., Vanderpas, J., Vanderpas-Rivera, M., Delange, F., and Ermans, A. M.,** Cassava products, HCN content and detoxification processes, in *Nutritional Factors Involved in the Goitrogenic action of Cassava,* Delange, F., Iteke, F. B., and Ermans, A. M., Eds., IDRC, Ottawa, IDRC-184 e, 1982, 51.

175. **Gomez, G., Valdivieso, M., Zapata, L. E., and Pardo, C.,** Technical note: cyanide elimination, chemical composition and evaluation in bread making of oven dried cassava peeled root chips or slices, *J. Food Technol.,* 19, 493, 1984.

176. **Cooke, R. D. and Coursey, D. G.,** Cassava: a major cyanide-containing food crop, in *Cyanide in Biology,* Vennesland, B., Conn, E. E., Knowles, C. J., Westley, J., and Wissing, F., Eds., Academic Press, New York, 1981, 93.

177. **Cooke, R. D. and Maduagwu, E. N.,** The effects of simple processing on the cyanide content of cassava chips, *J. Food Technol.,* 13, 299, 1978.

178. **Paula, R. D. G. and Rangel, J.,** HCN or the poison of bitter or sweet manioc, *R. Aliment.,* 3, 215, 1939.

179. **Charavanapavan, C.,** Studies on manioc and limabeans with special reference to their utilization as harmless food, *Trop. Agric.,* 100, 164, 1944.

180. **Tewe, O. O. and Maner, J. H.,** Long term and carry over effect of dietary inorganic cyanide (KCN) in the life cycle performance and metabolism of rats, *Toxicol. Appl. Pharmacol.,* 58, 1, 1981.

181. **Nambisan, B. and Sundaresan, S.,** Effect of Processing on the cyanoglucoside content of cassava, *J. Sci. Food Agric.,* 36, 1197, 1985.

182. **Fukuba, H., Igarashi, O., Briones, C. M., and Mendoza, E. M.,** Cyanogenic glucosides in cassava and cassava products. Determination and detoxification, in *Tropical Root Crops Postharvest Physiology and Processing,* Uritani, I. and Reyes, E. D., Eds., Japan Scientific Societies Press, Tokyo, 1984, 225.

183. **Bassir, O. and Fafunso, M.,** Effect of precooking processing on the cyanide contents of the leaves of eight ultivars of the cassava plant *Manihot esculenta, Plant Foods Man,* 2, 91, 1976.

184. **Collard, P. and Levi, S.,** A two-stage fermentation of cassava, *Nature (London),* 183, 620, 1959.

185. **Akinrele, I. A.,** Fermentation of Cassava, *J. Sci. Food Agric.,* 15, 589, 1964.

186. **Ekundayo, J.,** in *Fungal Biotechnology,* Smith, J. E., Berry, D. R., and Kristiansen, B., Eds., Academic Press, New York, 1980.

187. **Ikediobi, C. O. and Onyike, E.,** Linamarase activity and detoxification of cassava (*Manihot esculenta*) during fermentation for gari production, *Agric. Biol. Chem.,* 46, 1667, 1982.

188. **Ejiofor, M. A. N. and Okafor, N.,** Microbial breakdown of linamarin in fermenting cassava pulp, in *Proc. Triennial Symp. of the Inter. Soc. Trop. Root Crops-Africa Branch,* Terry, E. R., Doku, E. V., Arene, O. B., and Mahungu, N. M., Eds., IDRC, Ottawa, 1984, 105.

189. **Meuser, F. and Smolnik, H. D.,** Processing of cassava to gari and other foodstuffs, *Starke,* 32, 116, 1979.

190. **Zitnak, A.,** Assay method for hydrocyanic acid in plant tissues and their application in studies of cyanogenic glucosides in *Manihot esculenta,* in *Chronic Cassava Toxicity, Proc. Interdisciplinary Workshop,* Nestel, B. and MacIntyre, R., Eds., IDRC, Ottawa, IDRC-010 e, 1973, 89.

191. **Rosenthaler, L.,** in *Handbuch fur Pflanzenanlyse.,* Vol. 3, Klein, G., Ed., Springer-Verlag, Basel, 1932, 1042.

192. **Puiss, G.,** Untersuchungen zur Isolierung und Bestimmung von Blausaure in pflanzlichen Material, *Z. Anal. Chem.,* 190, 402, 1962.

193. **Hanssen, E. and Sturm, W.,** Quoted by Nahrstedt, A., Erb, N., and Zinsmeister, H. D., in *Cyanide in Biology,* Vennesland, B., Conn, E. E., Knowles, C. J., Westley, J., and Wissing, F., Eds., Academic Press, New York, 1981, 461.

194. **Eyjolfsson, R.,** Recent advances in the chemistry of cyanogenic glycosides, *Fortschr. Chem. Org. Naturst.,* 28, 74, 1970.

195. **Lehmann, G. and Zinsmeister, H. D.,** (Quoted by Nahrstedt, A., Erb, N., and Zinsmeister, H. D.), in *Cyanide in Biology,* Vennesland, B., Conn, E. E., Knowles, C. J., Westley, J., and Wissing F., Eds., Academic Press, New York, 1981, 461.

196. **Narley, F.,** Biosynthesis of cyanogenic glucosides in cassava (*Manihot* spp.), in *Chronic cassava Toxicity, Proc. Interdisciplinary Workshop,* Nestel, B. and Mac Intyre, R., Eds., IDRC, Ottawa, IDRC-010 e, 1973, 73.

197. **Montgomery, R. D.,** Cyanogens, in *Toxic Constituents of Plant Foodstuffs,* Liener, I. E., Ed., Academic Press, New York, 1969, 143.

198. **Ikediobi, C. O., Onyica, G. O. C., and Eluwah, C. E.,** A rapid and inexpensive assay for total cyanide in cassava (*Manihot esculenta* Crantz) and cassava products, *Agric. Biol. Chem.,* 44, 2803, 1980.

199. **Winkler, W. O.,** Report on hydrocyanic glucosides, *J. AOAC,* 34, 541, 1951.

200. **Mendoza, E. T., Kojima, M., Iwatisuki, N., Fukuba, H., and Uritani, I.,** Evaluation of some methods for the analysis of cyanide in cassava, in *Tropical Root Crops, Postharvest Physiology and Processing,* Uritani, I. and Reyes, E. D., Eds., Japan Scientific Soceites Press, Tokyo, 1984, 235.

201. **Gilchrist, D. G., Lueschen, W. E., and Hittle, C. N.,** Revised method for the preparation of standards in the sodium picrate assay of HCN, *Crop. Sci.,* 7, 267, 1967.

202. **Gillingham, J. T., Shirer, M. M., and Page, N. R.,** Evaluation of the Orion cyanide electrode for estimating the cyanide content of forage samples, *Agron. J.,* 61, 717, 1969.

203. **Epstein, J.,** Estimation of microquantities of cyanide, *Anal. Chem.,* 19, 272, 1947.

204. **Aldridge, W. N.,** The estimation of microquantities of cyanide and thiocyanate, *Analyst (London),* 70, 474, 1945.

205. **Asmus, E. and Garschagan, H.,** Use of barbituric acid for the photometric determination of cyanide and thiocyanate, *Z. Anal. Chem.,* 138, 414, 1953.
206. **Lambert, J. L., Ramaswamy, J., and Paukstelis, J. V.,** Stable reagents for the colorimetric determination of cyanide by modified konig reactions, *Anal. Chem.,* 47, 916, 1975.
207. **Nambisan, B. and Sundaresan, S.,** Spectrophotometric determination of cyanoglucosides in cassava, *J. AOAC,* 67, 641, 1984.

Chapter 4

CASSAVA SPOILAGE AND PRESERVATION

One of the major constraints in the utilization of cassava is the rapid perishability of the tubers after harvest. Biochemical changes and microbial infestation spoil the tubers and make them unfit for consumption. This very often poses problems as the transportation over very long distances to the processing sites leads to the deterioration in the quality of the tubers.

I. POSTHARVEST DETERIORATION OF CASSAVA TUBERS

Two types of deterioration have been reported in the case of cassava. The primary deterioration manifested as blue-black streaks which become more intensive towards the periphery of the cortex has been described.[1] Secondary deterioration is caused by invading pathogens. The vascular streaking phenomena is initially of a blue or blue-black color later turning brown in the form of vascular streaks which can be clearly seen in longitudinal sections of the roots.[2] The changes in color spread to parenchymal cells which turn to a bluish color.[3] The biochemical nature of spoilage of tubers have been investigated in detail[4] (Table 1). Rapid reduction in starch and moisture, and increases in dry matter and sugar content were observed during the course of deterioration of tubers after harvest. Involvement of enzymes like cellulases, amylases, and pectinases have also been recorded during the course of spoilage of cassava tubers. Both cellular and extracellular enzymes in harmony participated in the deterioration of cassava after harvest.[5]

A. Vascular Streaking in Cassava

The level of total phenols determined during the first 6 days of storage of fresh tubers in six varieties of cassava showed a decreasing trend during the 2nd day of storage.[6] During the 3rd day of storage a marked reduction in the phenol values was observed in the rind of the tubers (as shown in Figure 1). The rind of all varieties had a much higher phenol content when compared to the flesh portion. Most of the tubers developed vascular streaking in the form of blue-black lines along the periphery of the cortex of the tubers after 48 hr of storage. During the 4th day of storage, pathogen attack was initiated resulting in the rotting of the tubers. As rotting advanced to 60% in the tubers, the blue streaks were seen as brown streaks in the rotted area and a sudden up shoot of phenolic values was observed. As the rotting advanced to 74 to 80% and tissue degeneration set in, the phenol values were again lowered in all the varieties.

The titrable acidity status of the tubers maintained a uniform decrease up to the 3rd day of storage in all varieties. The rind of the tubers had a greater acidity as compared to the flesh. The sharp increase in acidity noticed as the pathogen invasion advanced in the tissue is shown in Figure 1. The increase in acidity coincided with the observed increase in phenols in the tubers stored for 5 days. As the rotting advanced, there was a decrease in acidity in the tubers and there was also pronounced decrease in phenols during this period.

The changes in the total phenol and titrable acidity during the postharvest storage of cassava seem to have a close association with the vascular-streaking phenomenon. For all the varieties tried, the vascular-streaking appeared during the 2nd or 3rd day of storage. It is possible that the polyphenolic compounds in cassava are oxidized to quinone-like substances which are complexed with small molecules such as amino acids to form colored pigments which in turn are deposited in the vascular bundles, thereby coloring the bundles.[7] Rickard has demonstrated by cytological studies and electron microscope observation that the discoloration occurrings in the physiological deterioration constitutes a damage or wound

Table 1
BIOCHEMICAL CHANGES DURING SPOILAGE OF FRESH CASSAVA TUBERS[4]

Constituents	Variety	Period of storage (days)							
		0	1	2	3	4	5	6	7
Dry matter (%)	H-1687	38.4	38.4	36.4	37.2	34.8	34.4	31.6	30.0
	H-2304	41.6	40.4	38.4	38.4	38.0	38.4	38.0	35.2
Starch (g/100 g fresh weight)	H-1687	28.3	27.3	24.0	24.0	24.7	22.7	23.3	23.3
	H-2304	29.2	29.2	28.3	24.7	24.7	24.0	24.0	24.0
Sugar (g/100 g fresh weight)	H-1687	0.70	0.80	0.89	0.80	0.95	1.05	1.10	1.31
	H-2304	0.59	0.59	0.65	0.87	0.95	1.00	1.05	1.05
HCN (mg/g dry weight)	H-1687	90	90	45	30	38	60	30	30
	H-2304	60	60	45	30	30	45	30	23

response by the root tissues, which does not remain localized but spreads rapidly through the root.[8,9] The initial response occurs as occlusions of the xylem vessels and the production of fluorescent compounds in the parenchyma tissues, and the occlusions contain carbohydrate, lipids, and compounds similar to lignin. In the initial stages free phenolic compounds, leucoanthocyanins, and catechins are found in the xylem vessels. This phenolic compound which exists in a conjugated form under normal conditions may be liberated as a result of injury.

Further biochemical tests showed many changes occurred in the phenolic constituents of the roots following injury. Positive surface test responses on cut-root pieces with vanillin, nitrous acid, and phloroglucinol — HCl reagents also demonstrated the presence of phenols and lignin-like material in the discolored root.

Spectrophotometric analysis of cassava root extracts demonstrated that considerable increase in total phenol, leucoanthocyanin, and flavanol contents were occurring in the root pieces following injury. Increase in phenylalanine ammonia activity, together with the increases in phenol content indicate that the latter changes are at least partly due to the novosynthesis and not only to qualitative changes in the phenolic constituents.

The development of colored material in the storage parenchyma and vascular tissue was observed to be accompanied by localized increases in the activity of polyphenol oxidase and peroxidase. It was concluded that some of the lignin-like cytochemical responses observed in the colored material were due to condensed tannins derived from leucoanthocyanidines and catechins. The compound with greatest fluorescence has been identified as scopoletin, a coumarin.[8,10] The presence of scopoletin in low concentration in the fresh tubers increases considerably (from 1 to 250 g/g dry matter) within 24 hr after harvest. Roots resistant to physiological deterioration were found to accumulate less scopoletin than susceptible roots.[10] The findings that vascular-streaking can be arrested by heating the tubers before storage and that cold storage prevents streak formation, give added support to the observation that the changes taking place during streaking can be an enzymatic oxidation and probably preheating or freezing may kill or inactivate the polyphenol oxidase activity.[3,11] Anaerobic atmosphere of CO_2, propane, etc., and oxygen depleted air, are known to inhibit the streak formation.[12] With oxygen being an essential substrate for polyphenol oxidase, under anaerobic conditions the enzyme cannot oxidize polyphenols. Mechanical injury of the harvested cassava tubers facilitates an accelerated entry of oxygen, thereby accelerating the vascular lining formation in the tubers.[13]

Effects of dipping the roots or tubers in solutions of some inhibitors/inactivators of polyphenol oxidase such as ascorbic acid, glutathione, or potassium cyanide at concentrations of 4×10^{-3}, 4×10^{-3}, and 5×10^{-3} M, respectively, for 8 hr indicate the possibility

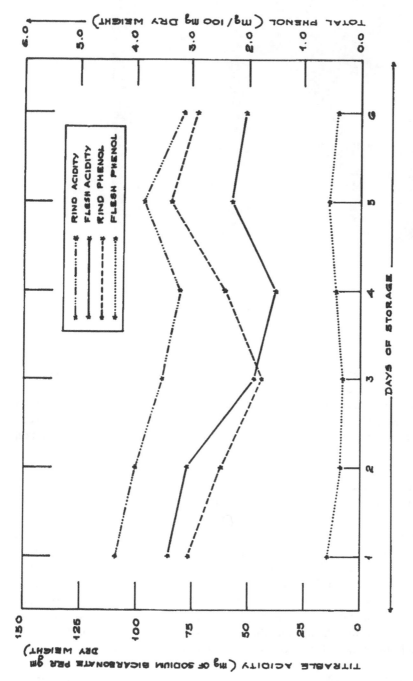

FIGURE 1. Titrable acidity and total phenol changes during the storage of cassava.

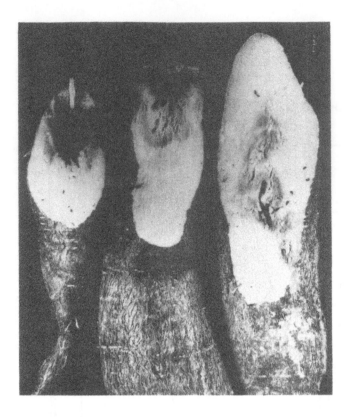

FIGURE 2.   Spoilage of cassava due to *R. oryzae.*

of polyphenol oxidase as a key enzyme associated with the blue vascular discoloration of cassava roots. In the treated roots, vascular discoloration was delayed, whereas these symptoms invariably occurred in the untreated roots after 2 days in a crude homogenate of cassava root polyphenol oxidase. Studies conducted on the effect of the above chemicals on a crude homogenate of cassava root polyphenol oxidase in vitro also supported their inhibitory action on the enzyme.[14]

B. Microbial Rotting of Cassava Tubers

Attempts made to isolate microorganisms associated with postharvest deterioration confirmed the predominant role played by the fungus *Rhizopus oryzae* (Figure 2). The fungus possesses mycelium of two kinds, one submerged in the substratum and the other aerial constituting stolons. These stolons are present on nodes, on which occur the rhizoides which are implanted in the substratum. The sporangiophores are single or in groups of two tiers or more. The sporangia, white at first, become bluish black at maturity, and are spherical or almost spherical-flattened at the base. Spores are found, oval, or angular in shape and are colored. The dry weight of the mycelium showed that varying concentrations of potassium cyanide when incorporated into the medium did not influence the growth of the fungus. *R. oryzae* was also able to utilize all the carbon sources provided in the medium, especially the phenolic substrate.[15]

A number of other fungi and bacteria have also been reported to be associated with the deterioration of cassava. They included *Rigidoporus liguosus, Lasiodiplodia theobromae, Rhizopus niger, Aspergillus flavus, Cylindrocarpon candidum, Trichoderma harzianum, Penicillium* sp., *Mucor* sp., *Fusarium* sp., *Glomerella* sp., *Gleosporium* sp., *Rhizoctonia* sp., *Bacillus* sp., *Xanthomonas* sp., *Erwinia* sp., and *Agrobacterium* sp.[1,11,16-18]

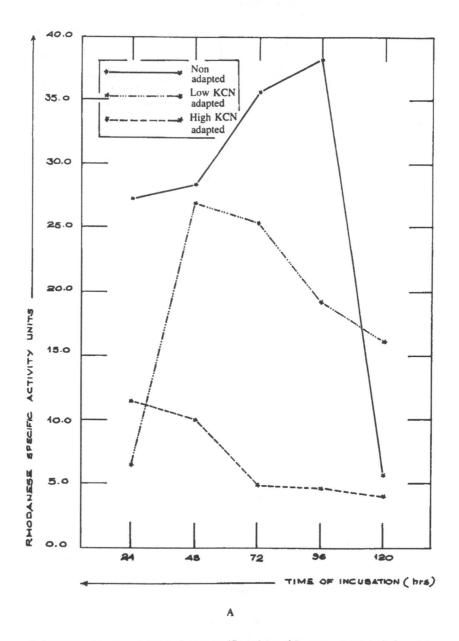

FIGURE 3.   (A, B, and C) Rhodanese specific activity of *R. oryzae* grown in 2, 3, and 4 m*M* KCN.

C. Cyanide Detoxification by *Rhizopus oryzae*

Degradation of cyanogenic glycosides of cassava by *R. oryzae* was studied by growing the organism in potato dextrose broth with and without linamarin and potassium cyanide.[19] The influence of the adaptability of the organism to low (1 m*M*) and high potassium cyanide (KCN) (5 m*M*) on the growth and release of rhodanese in a KCN-containing medium was also studied (Figure 3). Nonadapted spores of *R. oryzae* recorded poor growth as indicated by the mycelial dry weight when compared to the low and high KCN adapted spores. However, nonadapted spores released large quantities of rhodanese when compared to the adapted spores. The poor growth of *R. oryzae* in media containing linamarin indicated that linamarin was also toxic to the organism. However, the fungus has an efficient mechanism for detoxifying the cyanogenic glycosides and assimilating the cyanide in a utilizable form.

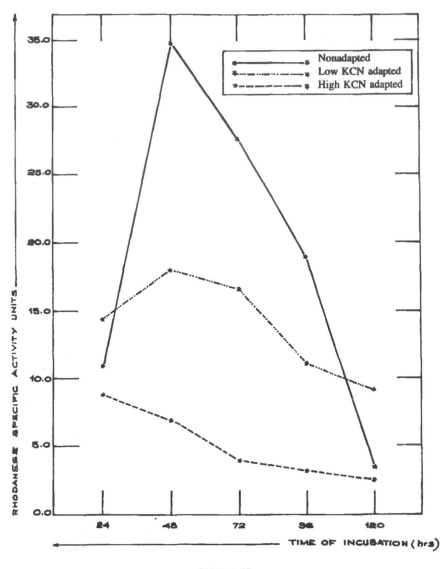

FIGURE 3B

Adaptation to cyanide seems to be a slow process and once detoxification is completed the fungus can grow rapidly. This explains the lag phase observed in the freshly harvested cassava tubers for the appearance of visual symptoms of fungal attack.

Enzymes which degrade polysaccharides exist in most parts of the higher plants and are readily released by most of the pathogenic bacteria and fungi. Some pathogenic bacteria are also capable of releasing degrading enzymes much more readily than even the pathogenic species. Cellulases and pectinolytic enzymes constitute the major group of degrading enzymes. In the degradation of starchy tissues, amylase plays a predominant role. The invading pathogens also release cellular enzymes mentioned earlier for the degradation of polysaccharides. Padmaja and Balagopalan have studied the cellular and extracellular enzymes associated with the spoilage of cassava.[5,20]

D. Cellular Enzymes Associated with Cassava Spoilage
1. Cellulase
Out of the five varieties of cassava tried, viz., H-165, H-226, H-1687, H-2304, and M-

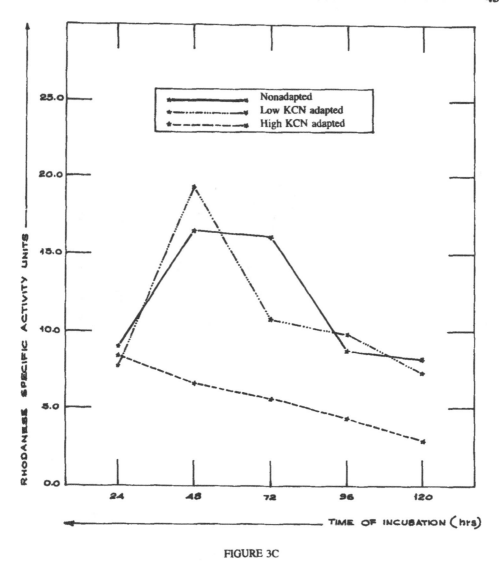

FIGURE 3C

4, the variety M-4 had the least activity in the fresh state. Highest activities were exhibited by H-1687 and H-226. There was a progressive increase in activity up to 6 days, thereafter a slight decrease was noted in most of the varieties (Figure 4).

2. Amylase

Peak enzyme activity was noticed on the 6th day of storage in the cassava of varieties H-226 and H-2304, whereas the activity was maximum on the 4th day in the caes of the other varieties as shown in (Figure 4). Decrease in starch content with concomitant increase in the free-sugar content during storage of cassava had been reported earlier.[4]

3. Polyphenol Oxidase

Considerable variation has been found in the concentration of the enzyme polyphenol oxidase in different varieties of cassava.[5,6,20] The level of polyphenol oxidase increased during the 2nd day of storage. This is followed by a decrease during the 3rd day but significant reduction in the rind polyphenol oxidase could be observed. The decreasing trend of enzyme activity continued during the 4th and 5th day of storage. Corresponding phenol concentration also decreased on the 4th and 5th day and this continued until vascular streaks appeared in the roots (Figure 1).

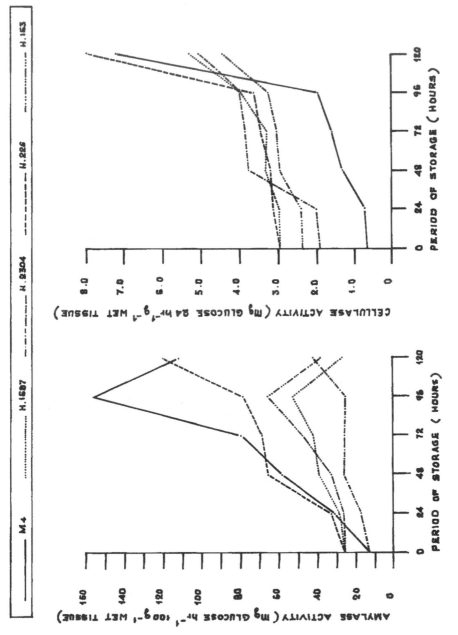

FIGURE 4.   Amylase and cellulase activity of cassava during storage.

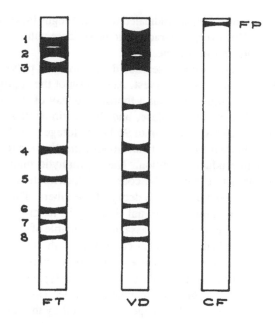

FT - FRESH TUBER
VD - VASCULAR DISCOLOURED TUBER
CF - CULTURE FILTRATE
FP - FUNGAL PEROXIDASE

1 TO 7 ISOENZYMES
8 MARKET DYE

FIGURE 5.   Gel electrophoretic band pattern of fresh and vascular discolored cassava tubers.

4. Gel Electrophoretic Studies

The gel electrophoretic band pattern indicated the presence of seven isoenzymes of peroxidase in fresh cassava tubers. The isoenzyme pattern of vascular discolored tubers (shown in Figure 5) indicated eight isoperoxidases with one additional band and increased intensity of staining.[20]

E. Extracellular Enzymes Associated with Cassava Spoilage

1. Amylase

There was a gradual increase in the amylase activity after 24 hr of incubation in the case of *R. oryzae* and peak activity was noticed at 96 hr. At 120 hr there was a sharp decline in the activity. The enzymes elaborated by *R. oryzae* were much lower than that of the endogenous enzyme.

2. Cellulase

Cellulase activity of *R. oryzae* showed a gradual increase for up to 48 hr and thereafter a gradual decline was noticed. This showed the rapid inactivation of the cellulases by *R. oryzae*. The pectin methylesterase and polygalacturonase activity of *R. oryzae* was found maximum at 48 hr and thereafter the activity started declining. During the initial stages of growth the polyphenol oxidase and peroxidase activity gradually started declining.

3. Polyphenol Oxidase and Peroxidase

A similar trend was observed in the case of polyphenol oxidase and peroxidase activity.

The electrophoretic pattern of *R. oryzae* indicated that the fungus lacks all the isoperoxidases present in cassava tubers. However, a characteristic peroxidase with very low electrophoretic mobility was present in the culture filtrate.

Though no visible symptoms were observed, degradation of polysaccharides and polyphenols was initiated immediately after harvest. Activation of the appropriate enzymes from the onset of harvest together brought about a thorough decay of the tubers. A progressive increase in the activity of amylase, cellulase, and the pectin-degrading enzymes could be observed in most of the cassava varieties up to 96 hr of storage.[5,6,19,20] The further decrease in the activity of these enzymes may be due to an enzyme inactivation resulting from substrate nonavailability and/or end-product inhibition. The saprophytic microflora residing in the harvested cassava become active during the course of deterioration, which along with the endogenous enzymes bring about a thorough decay of the tubers. This was evident from the behavior of *R. oryzae* which showed an initial time-lag in its growth. *R. oryzae*, the microorganism associated with the spoilage of cassava was capable of elaborating these enzymes into the growth medium and hence, can also contribute to the degradation of cassava polysaccharides in harmony with the host enzymes. The fungus plays only a trivial role in the vascular blue lining. The activity of peroxidase as well as polyphenol oxidase was minimum in the broth of the cultures after 24 hr of incubatin and decreased further, indicating that the role in blue lining is negligible. The meager role played by the fungus in vascular lining could also be understood from the polyacrylamide gel electrophoretic studies.[5] While there were seven isoenzymes of peroxidase in the fresh tubers and eight isoenzymes in vascular discolored tubers, only a feeble band with very low electrophoretic mobility was noticed in the culture filtrate of *R. oryzae*. This confirmed the major role of host peroxidases in the vascular blue lining phenomenon.

II. TECHNIQUES OF CASSAVA STORAGE

Many techniques have been reported to store cassava in its fresh state. The traditional methods consist of keeping cassava roots buried in the ground. In West Africa and in India⁺ surplus tubers are sometimes piled into heaps and watered daily to keep them fresh or coated with a paste of earth or mud which can preserve their freshness for periods of 4 to 6 days.[21] But there are few records of such storage under the farmers conditions.

A. Storage in Boxes

Fresh cassava could be satisfactorily stored in moist sawdust. The sawdust needed has to be moist but not wet. If too dry, the typical pattern or deterioration could occur in the roots, while if too wet, fungal and bacterial rot rapidly sets in.[22]

B. Storage in Plastic Bags or Film

The use of plastic bags or film wraps helps to reduce the loss of moisture in the tubers. Averre found that packing in polythene bags, within which the individual roots were wrapped in moist paper, was found to have some inhibiting effect on the development of vascular discoloration, the effect being more marked at a temperature below 10°C or above 40°C.[3] Techniques have been developed at CIAT, Cali, Colombia to prolong the shelf life of cassava by treating the roots with the thiabendazole-based fungicide and packaging them into polyethylene bags.[10]

Cassava roots are immersed for 5 min in solutions containing 0.4% mertect (Ciba-Geigy, Colombia) whose active ingredient is thiabendazole, 30 to 60 kg lots packed into sisal sacks are immersed in the fungicidal solutions. After treatment the roots are emptied onto a shaded area to dry for 15 to 30 min. This will help to drain excessive humidity. The sites of treatment should be as near as possible to the place of harvest to avoid physiological changes during transport. Undamaged roots are then packed in bags ranging from a capacity for 1 kg up to 20 kg of roots with the best results obtained from the small bag sizes (1 to 5 kg roots). Roots must be stored in a place offering protection from direct sunlight, rain, and rodents. In hot climates adequate ventilation to stop the internal bag temperature from exceeding 40°C is necessary to prevent roots from rotting. Roots must remain at least 4 days in an environment where curing can occur rapidly, therefore, immediate transport to cold climate areas is not advised since both physiological and microbial rotting can initiate from uncured areas of damage. Once cured, roots can be transported to any climate.

Bags should be closed until sold to consumers, the quantity per bag will thus depend on market requirements. Storage can continue after purchase by consumers, so long as bags are resealed after partial removal of contents to maintain the correct conditions inside the bag.

Averre[3] has reported successful storage of cassava roots under deep-freeze conditions in polythene bags. In many South American countries large quantities of tubers are preserved commercially under deep-freeze conditions.

C. Waxing

The effect of waxing cassava roots to improve their storage life was found to be successful in many countries under experimental conditions. Subramanyam and Mathur[23] have used a 2.2% aqueous emulsion of fungicidal wax, containing 17% triethanolamine and 5% O-phenylphenol in which tubers were dipped for 1 min, after which they were drained, dried, and stored at room temperature. The treatment could reduce the loss in weight. Dipping in wax without fungicides at a temperature of 90 to 95°C for 45 sec could prolong the shelf life of tubers up to 1 to 2 months.[23]

D. Chemical Treatments

Various methods for storage of cassava with chemical treatments have been reported. Selection of chemicals and estimation of residual toxicity are very important in such storage techniques. Treatment with an ethylene dibromide/ethyl bromide mixture gave a maximum storage life of 19 days, while steeping in a 1.2% formaldelyde solution for 30 min prior to storage in an airtight container had a storage life of 25 days.[24,25] Storage loss of cassava tubers could be reduced by dipping them in a 1% sodium O-phenyl phenate combined with 900 ppm 2:6 dichloro-4-nitroaniline for 45 sec followed by a water wash.[18] Storage life of cassava from a pruned plant was greatly increased by the use of manzate and sodium hypochlorite (4×10^3 and 2.5×10^4 ppm respectively) to prevent microbial rotting.[26] A stronger solution of manzate, 8×10^4 ppm could completely inhibit microbial rotting in tubers stored in polythene-lined bags. Treatment of roots with benomyl and chlorine water could also reduce the fungal and bacterial infections in cassava.

E. Field Clamp Storage

A method of storing cassava roots in structures similar to European potato clamps for periods of 1 to 3 months depending upon clamp design and prevailing ambient conditions was developed.[27] The basic design of these field clamps consists of a circular bed of straw or another material such as dried grass or dry sugar cane leaves (approximately 1.5 m in diameter and 15.0 cm thick after it has been compacted) is placed on suitable well-drained ground. The freshly harvested roots are heaped in a conical pile on this straw bed. The pile of roots is then covered with a similar layer of straw and the entire clamp covered with soil to a thickness of 15 cm. The soil is then removed from around the circumference of the clamp, forming a drainage ditch. In hot, dry conditions it was necessary to ensure that the internal clamp temperature does not exceed 40°C since roots deteriorate rapidly at a higher temperature. A thicker soil cover and the provision of ventilators to encourage airflow within the clamp are recommended to reduce the temperature inside the clamp.

FIGURE 6.   Storage of cassava in soil.

In very wet conditions, precautions need to be taken to prevent the roots from becoming wet within the clamp, since wet roots deteriorate rapidly. Likewise, roots that were rained upon during harvesting and handling should not be stored even if they were subsequently sundried. Wetting the soil cover during hot dry periods is also recommended for lowering the internal temperature of the clamp.

F. Storage in Moist Sand and Soil
Based on the basic information on the biochemical and microbiological causes for the postharvest deterioration of cassava, low-cost technology for extending the shelf life of fresh cassava tubers was developed at Central Tuber Crops Research Institute, Trivandrum, Kerala, India[28] (Figure 6). Undamaged bunches of freshly harvested tubers with a portion of the main stem are preferred for storage in moist sand/soil. After preparing pits of varying dimensions under shade, moist soil/sand is spread at the bottom.

The size of the pits usually depend upon the quantity of tubers to be stored. The moisture in the soil/sand is adjusted within the range of 10 to 15% by sprinkling water. Bunches of harvested tubers along with the intact stem could be arranged layer by layer, spreading the moist soil/sand in between the layers. After arranging three such layers the pits are to be covered with moist soil/sand. Since the top layer dries quickly, water may be sprinkled occasionally on the top layer.

Sprouting of the buds from the intact stem is very often noticed in the presence of moist soil/sand. The germinated buds from the bottom layer are unable to come out of the soil surface wile profuse growth of the apical buds from the top layer could be seen. Therefore, periodical nipping of the visible sprouts is required. While the undamaged tubers remain healthy throughout, damaged tubers very often deteriorate due to the activity of a bacterium suspected to be a *Bacillus* sp.

Fresh tubers could be stored in this manner for 2 months with little changes in the biochemical constituents (Tables 2 and 3). Though shelf life could be prolonged beyond 2

Table 2
BIOCHEMICAL CHANGES DURING STORAGE OF CASSAVA TUBERS IN SOIL PITS[28]

Constituent	Variety	Period of storage (weeks)								
		0	1	2	3	4	5	6	7	8
Dry matter (%)	H-1687	34.10	32.50	30.52	30.02	29.18	29.08	28.80	25.58	25.40
	M-4	42.80	39.60	38.35	37.95	36.65	35.90	34.20	33.10	32.46
	H-2304	45.70	43.00	42.40	42.16	41.98	37.76	38.60	—	—
Starch	H-1687	72.56	69.50	66.98	64.90	63.80	62.95	61.90	61.15	60.56
(g/100 g	M-4	76.85	74.16	71.15	69.85	68.68	67.10	64.85	64.78	63.20
dry weight)	H-2304	74.31	70.85	68.20	65.95	63.50	62.05	60.10	—	—
Sugar	H-1687	2.96	4.91	6.05	7.95	8.15	8.80	9.16	9.38	10.56
(g/100 g	M-4	1.50	3.50	5.50	6.85	7.58	8.08	8.08	0.10	10.48
dry weight)	H-2304	1.90	4.50	6.10	8.10	8.90	9.45	10.12	—	—
HCN	H-1687	143.56	138.22	120.57	98.59	70.48	50.98	46.64	40.65	34.32
(mg/g	M-4	101.02	93.80	90.98	62.03	59.95	43.60	31.80	25.58	19.40
dry weight)	H-2304	128.85	102.75	90.02	72.00	64.05	62.81	43.00	—	—

Table 3
BIOCHEMICAL CHANGES DURING STORAGE OF CASSAVA TUBERS IN SAND PITS[28]

Constituent	Variety	Period of storage (weeks)								
		0	1	2	3	4	5	6	7	8
Dry matter (%)	H-1687	33.89	33.00	32.98	30.16	30.08	29.80	28.65	28.40	28.85
	M-4	43.70	43.44	42.15	40.64	39.65	37.00	36.85	—	—
	H-2304	46.04	44.56	38.32	35.20	33.96	31.20	30.60	—	—
Starch	H-1687	70.37	68.25	67.68	65.42	64.90	63.18	62.10	61.16	60.48
(g/100 g	M-4	75.27	74.46	72.15	70.16	69.65	67.65	65.16	—	—
dry weight)	H-2304	72.31	68.98	67.95	66.28	64.38	62.16	61.28	—	—
Sugar	H-1687	2.51	3.84	4.05	5.91	6.65	7.50	8.55	9.09	10.12
(g/100 g	M-4	1.01	1.81	2.50	3.84	4.96	6.06	8.90	—	—
dry weight)	H-2304	1.33	4.14	5.66	6.08	7.42	8.80	9.38	—	—
HCN	H-1687	147.06	139.66	122.41	16.82	94.13	90.32	72.67	52.97	34.82
(mg/g	M-4	114.41	110.21	107.84	96.97	72.41	69.41	61.27	—	—
dry weight)	H-2304	122.28	133.64	110.58	95.30	73.53	54.82	32.89	—	—

months, such tubers become very sweet with poor cooking quality. However, these tubers could be utilized as cattle feed.

Moisture in the soil/sand prevented dehydration of the tubers and also controlled the aeration which could delay the enzymatic spoilage of tubers. Since the tubers used for storage were undamaged, there was no accessability for the entry of microorganisms in the tubers.

REFERENCES

1. **Booth, R. H.,** Storage of fresh cassava (*Manihot esculenta*). I. Post harvest deterioration and its control, *Exp. Agric.,* 12, 103, 1976.
2. **Montalco, A.,** Vascular streaking of cassava root tubers, *Trop. Sci.,* 15, 39, 1973.
3. **Averre, C. W.,** Vascular streaking of stored cassava roots, in *Proc. Ist. Int. Symp. Trop. Root Crops,* Trinidad, 2, 31, 1967.
4. **Maini, S. B. and Balagopal, C.,** Biochemical changes during post harvest deterioration of cassava, *J. Root Crops,* 31, 1978.
5. **Padmaja, G. and Balagopal, C.,** Cellular and extracellular enzymes associated with the post harvest deterioration of cassava tubers, *J. Food Sci. Technol.,* 22, 82, 1985.
6. **Padmaja, G., Balagopal, C., and Potty, V. P.,** Polyphenols and vascular streaking in cassava, *Cassava News Lett.,* 10, 5, 1982.
7. **Ramaswamy, S. and Rege, D. V.,** Cultivation of cassava and preparation of its products, *Mysore Agric. J.,* 27, 57, 1951.
8. **Rickard, J. E.,** Investigation into Post Harvest Behaviour of Cassava Roots and Their Response to Wounding, Ph.D. thesis, University of London, 1982.
9. **Rickard, J. E., Marriott, J., and Bahan, P. B.,** Occlusions in cassava xylem vessels associated with discolouration, *Ann. Bot.,* 43, 249, 1979.
10. **Wheatly, C. C.,** Storage of cassava roots for human consumption, *Cassava: Research, Production and Utilization,* UNDP and CIAT Publication, Colombia, 1985, 673.
11. **Singh, K. K. and Mathur, P. B.,** Cold storage of tapioca roots, Bulletin, Central Food Technology Research Institute, Mysore, India, 2, 181, 1953.
12. **Noon, R. A. and Booth, R. H.,** Nature of post harvest deterioration of cassava, *Trans. Br. Mycol. Soc.,* 69, 287, 1977.
13. **Marriott, J., Been, B. O., and Perkins, C.,** Etiology of vascular streaking in cassava roots after harvest: association with water loss from wounds, *Physiol. Plant.,* 44, 38, 1978.
14. **Padmaja, G., Balagopal, C., and Potty, V. P.,** Causes for the vascular streaking in cassava roots during postharvest deterioration, in *Proc. Sem. Postharvest Technol. Cassava,* AFST, Trivandrum, India, 1980, 17.
15. **Balagopalan, C., Maini, S. B., Potty, V. P., and Padmaja, G.,** Microbial rotting of cassava roots, in *Proc. Sem. Post Harvest Technol. Cassava,* AFST, Trivandrum, 1980, 23.
16. **Ekundayo, J. A. and Daniel, T. H.,** Cassava root and its control, *Trans. Br. Mycol. Soc.,* 61, 27, 1973.
17. **Wegman, K.,** Investigations on the causes of rapid spoilage of tapioca roots and the effectiveness of preservatives in tapioca flour, *Brot Gebaeck.,* 24, 175, 1970.
18. **Burton, C. L.,** Diseases of tropical vegetables in the Chicago market, *Trop. Agric. (Trinidad),* 47, 303, 1970.
19. **Padmaja, G. and Balagopalan, C.,** Cyanide degradation by *Rhizopus oryzae, Can. J. Microbiol.,* 31, 663, 1985.
20. **Padmaja, G., Balagopal, C., and Potty, V. P.,** Cellulolytic amylolytic and pectinolytic enzyme activities of deteriorating cassava roots, *J. Root Crops,* 8, 35, 1982.
21. **Rao, H. A. G.,** Cultivation of cassava and preparation of its products, *Mysore, Agric. J.,* 27, 57, 1951.
22. **Rickard, J. E. and Coursey, D. G.,** Cassava storage. I. Storage of fresh cassava roots, *Trop. Sci.,* 23, 1, 1981.
23. **Subramanyam, H. and Mathur, P. B.,** Effect of fungicidal wax coating on the storage behaviour of tapioca roots, Bulletin, Central Food Technology Research Institute, Mysore, India, 5, 110, 1956.
24. **Majumder, S. K.,** Some studies on the microbial root of tapioca, Bulletin, Central Food Technology Research Institute, Mysore, India, 4, 164, 1955.
25. **Majumder, S. K., Pingale, S. Y., Swaminathan, M., and Subramanyam, V.,** Control of spoilage in fresh tapioca tubers, Bulletin, Central Food Technology Research Institute, Mysore, India, 5, 108, 1956.
26. **Lozano, J. C., Cock, J. M., and Castano, J.,** New developments in cassava storage, in *Proc. Cassava Protection Workshop,* Brekelbaum, T., Bellotti, A., and Lozano, J. C., Eds., Centro Internacional de Agricultural Tropical, Cali, Colombia, 1977, 135.
27. **Booth, R. H.,** *Cassava Storage,* Bulletin, CIAT, Colombia, 18, 1975.
28. **Balagopal, C. and Padmaja, G.,** A simple technique to prolong the shelf life of cassava, *National Symp. Prod. Utilization of Trop. Tuber Crops,* Trivandrum, 1985.

Chapter 5

PRESERVATION OF DRIED CASSAVA PRODUCTS

In spite of the present day rapid transportation and distribution systems, an appreciable time lag occurs between the processing of harvested biomaterials and their consumption. This lag may vary from only a few hours to many months, during which time adverse changes in quality take place. The length of time a food product remains of acceptable quality when kept under a given environmental condition is commonly referred to as its shelf life, storage life, or keeping quality. Preservation of processed food involves minimizing serious losses in quality by preventing deterioration and spoilage of food. Deterioration is usually defined as involving those changes in quality induced by physical, chemical, and biochemical reactions taking place within the food itself, with or without the intervention of the physical environment (e.g., oxygen, carbon dioxide, water, light, heat, etc.). On the other hand, spoilage is defined as those changes in quality brought about by the action of biological agents, especially microorganisms and insects. Spoilage usually leads to more drastic changes in quality than deterioration.

I. PRESERVATION BY DEHYDRATION

Extreme perishability of cassava tubers has been the stimulus for the development of a range of processing techniques, resulting primarily in the production of dried products such as chips, pellets, flour, starch, sago, etc. Dehydration is one of the major means by which foods are processed and preserved. It offers an opportunity for eliminating spoilage problems. Preservation by dehydration is based on the fact that all microorganisms require a certain amount of moisture in order to grow. This requirement is generally expressed as water activity (a_w) which can be calculated by dividing the vapor pressure of a foodstuff by the vapor pressure of pure water while both are at the same temperature. In general, bacteria have a higher a_w than yeasts and yeasts have higher a_w than molds. Accordingly, at a_w values of less than 0.70 foods are not likely to be spoiled by microorganisms at all. This a_w value is reached in dried vegetables at 14 to 20% moisture and in flour at 13% moisture.[1]

Deterioration of cassava products during storage, in a manner similar to any other dried produce, is mainly caused by exogenous factors such as fungi and insects. In view of this, adequate initial drying of any cassava product to a moisture content in equilibrium with 65 to 70% relative humidity and subsequent maintenance of this low moisture content are the fundamental prerequisites of satisfactory storage.[2] Susceptibility to damage by storage insects and molds is all the more increased by the markedly hydroscopic nature of most cassava products,[3] a clearly disadvantageous property considering the high humidity environments of most cassava-producing countries.

A. Packaging and Storage

Suitable packaging and storage facilities have to be provided to offer continued protection against biochemical deterioration, microbial growth, and insect infestation[4] which are likely to arise due to any rehydration by the uptake of atmospheric oxygen. Water vapor transmission resistance (WVTR) is frequently the principal consideration in protective packaging of cassava products.

Unlike cereal grains which are well suited to handling and storage in bulk, dried foodstuffs are normally packaged in some way for storage and marketing. The size of the package, however, may be large or small depending upon whether it is meant for a trade customer or a household consumer. Besides protecting the foodstuffs from mechanical hazards (shock, vibration, and compression) during transport, and from climatic (moisture, temperature, light, and oxygen) as well as biological (insects and microbes) hazards during storage, packaging also serves in the present day context as a vehicle for promotion and advertising of the product and consumer information. This aspect of the packaging can be of considerable importance in case of fast and convenience foods developed from cassava, viz., sago, semolina, macaroni, vermicelli, etc.

II. CASSAVA CHIPS

A. Magnitude of Postharvest Losses

Stored cassava chips are readily attacked by insects, resulting in quantitative as well as qualitative losses. The magnitude of postharvest losses has been reported to be as high as 16% by weight in Malaysia after 2 months of storage of cassava chips.[5] It has been reported in India that parboiled chips could be stored for 9 months with only a 3% loss in weight and with a 4 to 5% loss in an ordinary room.[6] However, the same study states losses of 10 to 12% in arah and 12 to 14% in the room for plain sundried chips. In another study conducted recently at CTCRI,*[7] the amount of dust formed ranged from 4% to 15 due to infestation when cassava chips were stored in paper bags for 120 days. The main factor in bringing about the actual loss of weight during storage of dried cassava chips is consumption by the major insect pests.

Dried root pieces of cassava are often riddled with holes by boring insects. Chips with increased surface area are more liable to insect infestation.

1. Insects Infesting Dried Cassava

A wide-range of species that feed directly on the dried foodstuffs and some mold-feeding insects have been reported as causes of weight loss in stored dried cassava chips. In Malaysia, *Rhyzopertha dominica* was observed as the most destructive insect in the well-dried cassava chips, while *Lasioderma serricorne* and *Araecerus fasciculatus* also were noted as serious pests of dry cassava chips.[5]

Ingram and Humphries[8] in a review of cassava storage listed 15 insect species found infesting cassava chips, of which 6 were stated to be major pests, viz., *Ahasverus advena*, *Araecerus fasciculatus, Rhizopertha dominica, Sitophilus oryzae, Stegobium paniceum*, and *Tribolium castaneum*. In a study conducted in India,[4] in freshly processed cassava chips stored in various conventional packaging materials, *A. fasciculatus* was found to be the commonest insect with the population of adult insects ranging from 7/kg cassava in the case of a metal container to 243/kg in a jute bag, at the end of a storage period of 120 days.

In studies made to ascertain the relative susceptibility of chips prepared from different cassava varieties, Pillai observed some resistant varieties on the basis of insect infestation and damage by *A. fasciculatus*, the most important among storage pests of cassava chips in India.[9]

B. Insect Control Measures

When rapidly dried to a moisture content of 13% or below and stored in sealed containers, good quality cassava chips keep well, but insect infestation is common when chips are stored in bulk or in open containers such as jute bags, particularly if they are kept for 6 months or longer.[8] Parker and Booth are of the opinion that infestation of chips occurs during the sundrying process and that insect damage during storage of cassava chips may be reduced by producing high quality rapidly dried chips and by storing them away from major sources of infestation, particularly other stored cassava chips.[5] Their observation that after 2 months of storage large numbers of insects were found in chips that had been dried slowly in the

* Central Tuber Crops Research Institute.

sun but oven-dried chips were insect free, indicates the advantage of artificial drying for chips meant to be stored. Alternatively, use of 3% salt (NaCl) mix as a preservative followed by slow sundrying also tends to reduce subsequent insect damage.[5,10]

1. Improved Storage Management

The extent of storage deterioration of primary dried cassava products is greatly affected by storage management.[5] Adequate drying for storage, attention to basic storage hygiene (such as sealing the cracks and gaps in the walls and floor of the storage structure or warehouse, avoiding darkness and dampness, placing dunnage and building uniform sized stacks, whitewashing where necessary, etc.) and the avoidance of prolonged storage under humid conditions are all practicable and necessary means of preventing, or at least reducing, deterioration.

These simple steps that could help to maintain the dried products in storage are often ignored and attempts are made to concentrate on chemical methods of insect control. However, in the absence of sanitary warehousing practices, chemical methods are rendered less effective.

2. Protective Packaging

Cassava chips, as many other dried cassava products, are a low-cost high-volume commodity. Manufacturers suggest that there is little justification for the financial investment required for sophisticated storage structures and techniques to reduce the losses even though adequate technology exists to do so. In this regard, some of the conventional packaging materials can offer good protection against the chemical deterioration and microbial attack brought about by changes in the humidity of the environment and also to a certain degree from external insect attack. Polyethylene film, laminated gunny sacks, laminated bitumin and polyethylene film, or high density polyethylene or polypropylene lined wooden sacks can be used to retard the effect of moisture loss in the cassava chips during storage.[11] In a study conducted to evaluate the conventional packaging materials for storage of cassave chips on the basis of change in moisture content, population of infesting insects, amount of dust formed as well as quantity of infested chips, metal containers and polyethylene-lined jute bags have been found to be the most effective packaging materials for safe storage of cassava chips for up to 90 days.[4]

3. Chemical Methods

Phosphine and methyl bromide are fumigants widely accepted as effective and safe that should be recommended for the disinfestation of dried cassava products in bulk storage. For small-scale treatments, less penetrating liquid fumigants such as ethylene dibromide may be effective.[2] The fumigation of bagged infested chips using methyl bromide, ethylene dibromide, or a mixture of ethylene dichloloride and carbon tetrachloride has been found to provide effective control of infestation. Methyl bromide and ethylene dibromide residues in the chips were found to be of the order of 3 to 15 ppm.[11]

The use of approved insecticides in periodic sprays applied to the surface of large piles or bulk bins of cassava chips, if applied soon after chipping and drying, has also proved to be effective.[2] Freshly dried cassava chips are unlikely to carry internal insect infestation, and subsequent infestation by primary and secondary pests, tend to be located mostly in the outer layers of a stack.[2] In this respect, infestation of cassava chips differs from cereal grains which are likely to carry primary infestation dispersed throughout the grain mass.

Other control measures that have been suggested are gammexane smoke bombs, fogging with DDT, and disinfection of warehouses before storage.[11] Jute bags impregnated with a lindane/dieldrin mixture have been found to prevent infestation of stored cassava chips from external sources, but do not control any existing infestation.[12]

C. Microbial Spoilage of Cassava Chips

Development of bacteria and fungi is quite often the first visible sign of spoilage in dried cassava products. Actual losses due to such attacks are not normally very high, but fungal growth in particular may attract mold-feeding insects and thus lead to increased total infestation. In some cases, fungi or bacteria can present a severe health hazard.

Material loss due to fungal or microbial infection of cassava products seems to be brought about by some conversion of the sugars and starch into CO_2 and water as a result of the metabolic activities of the microorganisms. Losses in small pieces of fresh cassava, of 62% moisture content, ranged from 6.2 to 58.3% after 10 days, and from 13.3 to 73.5% after 20 days by deliberate infection with various *Aspergillus* sp. during the drying phase.[13] These results may not have relevance to practical storage situations but are indicative of the massive losses of dry matter that even fungi can produce when chips or other cassava products are not adequately dried or are allowed to rehydrate to a serious degree.

MacFarlane,[2] in an extensive review, has listed the regular occurrence of 26 fungal species on chips or pieces of cassava notably *Aspergillus* sp. and *Penicillia* from Ghana as well as from Zaire and India. Infection of cassava chips by a wide variety of bacteria, fungi, and yeasts have been reported in India.[14] Cassava chip samples from 9 different markets of Kerala State in India have been found to harbor bacterial population ranging from 1.13×10^4 to 55.55×10^4/ g and fungal population ranging from 0.17×10^3 to 1.57×10^3/ g.[15] *Rhizopus, Penicillium, Aspergillus, Fusarium,* and yeasts were the predominant fungal genera observed on different market chips.

As the fungi *Aspergillus* in general, and, in particular, *A. flavus* are the main storage fungi in dried cassava products, production of aflatoxin can also be expected. In fact many of the *A. flavus* strains are capable of producing aflatoxins when grown in vitro on cassava.[2]

1. Control Measures

The primary requirement for fungal growth is inadequate drying or rehydration of the initially dried chips. But it has been observed that *A. flavus* and other fungal species often develop on cassava chips or slices during the drying process. Proper drying of cassava products, therefore, assumes critical importance in their subsequent storage behavior.

Good results in preventing mold development have been obtained by steeping chips in sulfurous acid and storing them in polyethylene bags.[2] Sodium chloride (3% mix) or sodium metabisulfite (5% mix) in the predrying storage of wet cassava chips are both capable of maintaining good color and texture, and preventing gross microbial breakdown for up to 4 weeks and also during slow sundrying under adverse weather conditions.[10]

III. CASSAVA FLOUR

A. Magnitude of Spoilage

Cassava flour or meal is usually the final product obtained by pounding or grinding the dried chips. It is usually regarded as a difficult product to preserve in storage. Cassava flour is more hygroscopic than chips and starch.[7] Probably faster reabsorption of atmospheric moisture by the tiny particles is responsible for this. However, sometimes flour is preferred to storage to chips because besides a relatively higher bulk density and consequently lesser demand for space, some insect species, especially borers, which attack dried chips do not thrive in flour.

Cassava flour rapidly adjusts to the humidity of the surrounding atmosphere. It has been reported that flour stored at 6.7% moisture content took up an additional 5% in 6 months when packed in cloth or gunny bags, but only 1 to 2 % in polythene-lined sacks.[3]

Total weight loss in exports of cassava meal from Tanzania have been reported to be 10 to 12%.[2] During storage in various conventional packaging materials, a maximum weight loss of only 4.8% has been found after a period of 120 days from experiments conducted in India.[7]

B. Insect Infestation of Cassava Flour

Although cassava flour is likely to be infested by a wide range of insect species, *T. castaneum* has been the most common species.[2,7,8]

1. Control Measures

Recommendations for insect control have included the use of gammexane smoke generators, DDT fogging, and disinfestation of warehouses before storage, but caution is needed in the choice of insecticides due to their hazardous nature.[2] In a comparative evaluation of conventional packaging materials for storage of cassava flour, polyethylene-lined jute bags were found only to be most effective for safe storage of flour up to 90 days.[16]

C. Microbial Spoilage

In samples of cassava flour and meal, mainly of Thai origin and from some other parts of the world, bacterial counts had been observed to range from 1 million to 185 million/g. The principal bacteria identified were *Bacillus* and *Enterococci* while of the fungi, quite a few *Aspergillus* species were detected.[2] Proper drying and water-vapor transmission resistant packaging during storage are the invariable measures to be adopted to prevent fungal or bacterial infection.

IV. PELLETS

Most of the dried cassava for export to the developed countries is now handled as pellets rather than chips or flour. Quite often pelleting has to be carried out at a critical moisture content which is high from a storage point of view. Cooling and drying is then necessary before subsequent storage and handling of the pellets. But the hard glossy pellets could be stored without spoilage for 4 months in plastic bags. Hard and dense pellets are expected to be somewhat less susceptible to insect infestation. For protection of exported pellets, incorporating a suitable insecticide at an acceptable dosage rate in the binder required for the consolidation of pelletized cassava meal is a conceivable procedure. Such treatments can substantially reduce the need for fumigation with economically justifiable extra cost where, of course, no legal impediment exists.[2]

V. STARCH

Cassava starch, being powder in form and hygroscopic in nature, is prone to pickup moisture at a faster rate.[4] Experiments with cassava starch stored in a modified atmosphere showed that the equilibrium moisture content is attained in 15 days. The critical moisture-molding content of cassava starch was found to be 19%.[16] It was, therefore, recommended that cassava starch be stored well below 10% moisture to keep it free from mold growth.

A graphic record of variation in the moisture content of the cassava starch stored in different conventional packaging materials is given in Figure 1, along with the changes in average ambient temperature and relative humidity in the vicinity of a storage site under a typical humid climate.[4] As the relative humidity increases there is a corresponding increase in the moisture content of starch stored in packaging materials that are not effective moisture barriers (viz., jute bag and breached low density polyethylene LDPE bag). However, such changes are not reflected immediately but with a lag that corresponds to the time taken for attaining equilibrium, as has been noted earlier.

In this study, high density polyethylene (HDPE) bags and polyethylene-impregnated jute

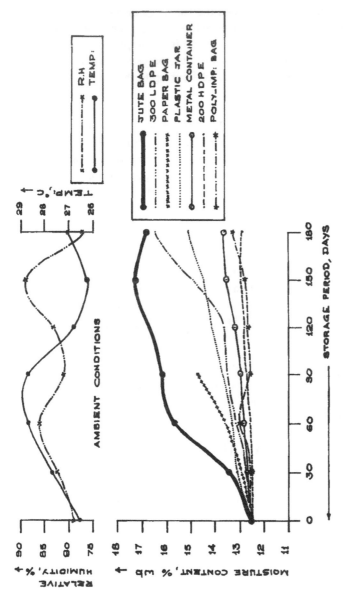

FIGURE 1.   Variation in moisture content of cassava starch stored in different packaging systems.

bags have been recommended for storing cassava starch up to 180 days on the basis of determining moisture, viscosity, total microbial count, and viable bacterial and fungal populations.

The literature records no problems of insect infestation in stored starch.[2] It is considered that pure starch is not relished by the insects.[4]

VI. SAGO

Sago (Saboodana) or pearl is made from starch cake, globulated, and roasted. It presents little storage problems provided the usual protection is given against moisture and infestation. Generally white in color, it darkens with the onset of microbial growth. Moisture gain results in softening and moisture loss causes brittleness in it. Both steamed and roasted tapioca grains store well in a variety of containers for 12 months or more and are generally insect resistant.[2]

The moisture absorption study on sago carried out at 27°C has indicated that the product with an initial moisture content of 9% (corresponding to 10.5% on dry weight basis) equilibrated to 41% whereas at 11% moisture, the equilibrium relative humidity was 53%.[11]

VII. VERMICELLI, SPAGHETTI, AND MACARONI

Vermicelli and spaghetti are made from wheat semolina or flour, cassava flour, and other ingredients. Both are in the shape of long strands, but vermicelli is much thinner. As these contain about 78 to 80% starch and 10% protein, they are predominantly moisture sensitive. These products reportedly have two critical moisture contents, the upper limit corresponding to softening and susceptibility to fungal spoilage, and the lower limit at which they become brittle and fragile. A suitable package has to protect the contents so that the moisture content lies between these two limits during distribution and storage, and also against physical damage as otherwise breakage of the product may lead to rejection by consumers.

Research studies on vermicelli have indicated that at 27°C, the product having a moisture content of 12% equilibrated to 70% relative humidity whereas at 10% moisture, the equilibrium relative humidity was 50%. Moisture contents of 13.2 and 9.5% both on dry weight basis, were found to be the upper and lower critical values for the product.[11]

HDPE pouches of 200 gauge and LDPE pouches of 300 gauge have been suggested for packaging vermicelli based on an estimated shelf life of 83 days under accelerated conditions, 38°C temperature and 92% relative humidity.

Macaroni products which are usually tube-like in form, are available in various sizes and shapes. The CFTRI, Mysore, India had developed a process for tapioca-macaroni using a blend of 60 parts cassava flour, 15 parts low fat groundnut flour and 25 parts wheat semolina.[12] Moisture absorption studies on tapioca-macaroni have indicated that the product has more protein absorbed and less moisture at all humidities, as compared to the predominantly starch-containing sago and vermicelli.[11] The shelf life of macaroni is, as a result, greater than that of other products since it is generally manufactured to a low moisture level of 7.5 to 8%. Storage trials covering a period of 1 year on both steamed and roasted tapioca-macaroni, and nutro-macaroni packed in different containers, have shown that all three products properly stored for over 12 months in normal conditions were also fairly resistant to insect attack.[2]

REFERENCES

1. **Stewart, G. F. and Amerine, M. A.,** *Introduction to Food Science and Technology,* Academic Press, New York, 1973, chap. 5.
2. **McFarlane, J. A.,** Cassava storage. II. Storage of dried cassava products, *Trop. Sci.,* 24, 205, 1982.
3. **Kuppuswamy, S.,** Studies on the dehydration of tapioca, *Food Sci.,* 11, 99, 1961.
4. **Nanda, S. K.,** Evaluation of packaging materials for storage of cassava starch, *J. Root Crops,* 11, 45, 1985.
5. **Parker, B. L. and Booth, R. H.,** Storage of cassava chips (*Manihot esculenta*): insect infestation and damage, *Exp. Agric.,* 15, 145, 1979.
6. **Hirandani, G. J. and Advani, K. H.,** Report on the Marketing of Tapioca in India, Marketing Ser. No. 88, Directorate of Marketing and Inspection, Ministry of Food and Agriculture, India, 1955, 72.
7. **Anon.,** *CTCRI Annual Report 1985,* Central Tuber Crops Research Institute, Trivandrum, India, 1986, 40.
8. **Ingram, J. S. and Humphries, J. R. O.,** Cassava storage: a review, *Trop. Sci.,* 14, 131, 1972.
9. **Pillai, K. S.,** Susceptibility of cassava chips to *Araecerus fasciculatus,* in *Proc. 4th Symp. Int. Soc. Trop. Root Crops.,* Cock, J., MacIntyre, R.,and Graham, M., Eds., IDRC, Ottawa, 1977, 202.
10. **Booth, R. H. and Dhiauddin, M. N.,** Storage of fresh cassava (*Manihot esculenta*). III. Preserving chipped roots before and during sun-drying, *Exp. Agric.,* 15, 135, 1979.
11. **Kumar, K. R. and Anandaswamy, B.,** Packaging and storage of dry tapioca products, in *Proc. Sem. Post Harvest Tech. Cassava,* Rajaraman, K. et al., Eds., Association of Food Scientists and Technologists, Trivandrum, India, 1981, 28.
12. **Anon.,** *The Wealth of India, Raw Materials,* Vol. 6, Council of Scientific and Industrial Research, New Delhi, 1962, 293.
13. **Clerk, G. C. and Caurie, M.,** Biochemical changes caused by some *Aspergillus* species in root tuber of (*Manihot esculenta* crantz), *Trop. Sci.,* 10, 149, 1968.
14. **Balagopal, C., Bhavani Devi, S., and Pillai, K. B.,** Microflora of cassava chips, *Agric. Res. J. Kerala,* 12, 80, 1874.
15. **Balagopal, C. and Nair, P. G.,** Studies on the microflora of market cassava chips of Kerala, *J. Root Crops,* 2, 57, 1976.
16. **Etorma, S. B.,** Chemical Studies on cassava products. I. The critical moisture molding content of cassava starch, *Philipp. J. Agric.,* 7, 409, 1936.

Chapter 6

PROCESSING OF CASSAVA

Cassava is extremely perishable and must be either consumed or processed within a few days of harvest. A study conducted by the American Academy of Science has indicated that about more than 25% of the cassava production of the world is lost in postharvest handling.[1] In lesser developed countries, the postharvest losses of cassava tubers are estimated to be 10%.[2] Cassava is also bulky and heavy, and hence expensive to transport over long distances. As a result, the crop is largely consumed in the vicinity of the production site, i.e., rural areas, and has not greatly penetrated large urban markets. In the past, India and elsewhere have not paid much attention to the development of postharvest technology of cassava, probably because most of the production is under subsistence farming conditions. Of late, its importance has been realized, not only as high energy food but also as industrial raw material and animal feed mix. This chapter describes the current practices and the recent research on more efficient methods for the major unit operations in the primary processing of cassava, viz., chipping, pelleting, and drying.

I. CASSAVA CHIPS

As fresh cassava tubers deteriorate within a few days after harvest, they are sliced into small pieces or chips and dried in the sun. The dried chips can be preserved for months and consumed after grinding into flour. They are also used in the industry for starch, dextrin, and glucose production. Chips are produced extensively in Thailand, Malaysia, Indonesia, India, and some parts of Africa.[3] Chips are produced in various forms, sizes, and shapes in different parts of the tropical world and known by different names, e.g., kokonte, gaplek, bombo, cossottes, etc.[4,5]

Edible tapioca chips in India are generally of two types. White chips are obtained by removing the outer skin of the tubers, slicing, and sundrying. Parboiled chips are obtained by immersing the chips in boiling water for 10 min before drying.[6] Parboiled chips are easier to store than white chips.

There is not much uniformity in the preparation of cassava chips in different localities or in Kerala, the largest cassava-producing state of India. The various chips available in the market can be grouped into the following categories although names may differ in other regions.

- Iritty type — outer skin is removed
- Vella — both rind and skin are removed
- Chilta — both rind and skin are retained

Vella chips usually attain a better price when compared to the other chips because of their bright color. The chilta and iritty are used for the cattle feed industry.

A. Mechanical Devices to Prepare Cassava Chips

Under conventional practices, cassava tubers are sliced with the help of hand-knives with or without peeling the outer skin and rind (Figure 1). Chips are then dried under the sun. However, this method is tedious and time consuming. Besides, it leads to uneven and delayed drying as cassava chips are produced in various forms, sizes, and shapes at different places. For the large-scale preparation of cassava chips, manually operated and motor-driven chipping machines have been developed in various countries.

FIGURE 1.   Traditional cassava chipping with handknives in India.

1. Hand-Operated Cassava Chipping Machine

A low-cost hand-operated cassava chipping machine developed at the Central Tuber Crops Research Institute (CTCRI) Trivandrum, Kerala, India,[7] seems to be a promising one (Figure 2). The machine consists of two concentric mild-steel drums, with the annular space divided into compartments for feeding the tuber, and the machine is supported on four legs. A rotating horizontal disk at the bottom of the drum carries the knives assembly. A pair of bevel gears is provided to operate the machine manually from the top and the chips are collected at the bottom. This machine produces round chips. The thickness of the chips produced can also be adjusted in steps, if desired.

The overall dimensions of the machine are 700 × 580 × 680 mm with a total weight of only 33 kg. Average hourly turnout of this machine is about 62 kg/hr for producing 5 mm thick chips as compared to a turnout of about 24 kg/hr by hand-slicing for producing 12.5 mm thick chips (Table 1). However, the machine results in 9 to 10% breakage which includes broken chips as well as shreds. Field evaluation studies conducted in different villages showed the average rate of adaptability of this machine among farmers to be 81.2%[8]

2. Malaysian Chipping Machine

The cassava chipping machine used in Malaysia (Figure 3) consists of a rotating circular steel plate of about 12 mm thickness and 100 cm diameter,[9] to which 4 or 6 blades are attached. The blade consists of 1 to 1.5 mm steel plate that is corrugated at the cutting edge. The machine produces strips about 6 mm wide and 3.6 mm thick, with a small percentage of fine particles. The length of the strips is dependent on the position of the roots at the time of cutting. The rotary disk is vertically mounted on the driving shaft and driven by a 3 bhp gasoline engine or electric motor with a belt-drive mechanism at a required disk speed of 500 rpm. This chipper has a turnout of 1 t/hr when operated by two men and fed manually. Since the chips produced from the Malaysian chipping machine are finer in size, the drying process becomes faster and the pounding and pelleting operations are easier.

FIGURE 2. Hand-operated cassava chipping machine developed at CTCRI.

Table 1
RATED CAPACITY FOR THE CTCRI CASSAVA
CHIPPING MACHINE AND MANUAL CHIPPING

Average chip thickness (mm)	Intermittent operation (single operator) Average turnout (kg/hr/operator)	Continuous operation (two operators) Average turnout	
		(kg/hr)	(kg/hr/operator)
2.3	28.1	38.4	19.2
4.6	61.8	97.2	48.6
6.9	73.3	117.8	58.9
(Hand-sliced) 12.5	24.2	—	—

Note: Rated capacity is on a fresh weight basis.

3. Thai Chipping Machine

The cassava chipper used in Thailand[9] consists of a thin circular plate instead of a thick heavy plate as used in Malaysia. The plate, which is normally made out of the cover of an oil drum, is cut and formed to produce cutting elements throughout its surface. This rotating notched disc is mounted on a wooden frame equipped with a hopper. The chips or shreds

FIGURE 3.   Malaysian cassava chipping machine.

produced are irregular in shape and take usually 2 to 3 days to reach a moisture content of 13 to 15% when dried on a concrete surface with periodic turning of the chips.[10]

4. Cassava Chipping Machine in Nigeria

The machine consists of an assembly of three knives held in slots of a metallic cylinder in a fashion similar to a wood planer.[11] Roots cut to 15 to 20 cm lengths are fed into a hopper placed above the horizontally mounted cylinder. The knives assembly rotates clockwise at 375 rpm, belt-driven by a 1 hp electric motor, while another cylinder of smaller diameter placed parallel at a gap of 2.5 mm from it rotates counter-clockwise at 95 rpm. The machine is powered by a 1 hp electric motor. Chips, removed progressively from the roots by the knives are flung below through the gap between the cylinder and are guided down by the bottom part of the hopper. Chips of 30 to 75 mm length are produced by this machine, with finer particles ranging up to 8%. Average turnout of this machine is reported to be 225 kg/hr.

Besides the above, some pedal-operated bicycle and treadle types as well as hand-operated chipper-cum-grater units developed in the Philippines,[12] can be suitable for the small-scale cassava processors and farmers.

II. CASSAVA PELLETS

Thailand and Indonesia are the world's largest exporters of dried cassava, largely in the form of pellets. The recent impetus to pelleting has come from the demand generated within the European Economic Community (EEC) for cassava products as a source of energy in compound animal feed.[13] Cassava pellets were introduced in order to make handling easier and also to maintain uniformity in size and shape. Another point in favor of pellets is the reduced volume-to-weight ratio which helps lower the shipping costs.

Pelleting is a process of particle size upgrading, utilizing a principle of compaction and extrusion. In the process a loose bulky material, ranging from a fine powder to small granules, is compressed and formed into a pellet of increased bulk density in a variety of sizes.

A. Cassava Pelletizing Process

Two types of presses are in operation in the pelletizing plants used in Thailand and Indonesia.[14] One press uses a horizontal circular plate die with a large number of holes drilled to form a honeycomb pattern. A set of six rolls are resting on the die which rotate by friction when the die is rotated around a vertical axis. The other type of press uses a vertical ring die with a set of two rolls inside the ring. Holes are drilled radially throughout the face of the die. When the ring is rotated, the rolls also rotate with it by friction.

Chips are first screened to remove sand and soil, and to separate the oversized ones, which are passed through a hammermill. Cassava particles are then conditioned to the correct moisture content with the use of a water spray and fed in front of the press rolls, which crush and force the material through the holes. As the compressed material emerges from the other side of the die, a knife shears the pellets to the adjusted length. Cassava pellets are usually less than 1 cm in diameter and 2 cm in length.[10]

The pellets from the presses are quite warm and soft. These are sent to coolers where the temperature and moisture content are reduced, after which the pellets become reasonably hard. The air-cooled pellets are sent to a screen from which the oversized are sent to a bagging machine or storage, and the undersized are returned to the feeding system. Proper cooling of the pellets reduces the moisture content by about 3 to 4%.[9,13]

Two types of pellets are recognized. "Brand" pellets are produced on imported European machinery. "Native" pellets are made on locally manufactured pelletizers with cheaper dies and rollers. Some quality-conscious feed industries in the EEC consider native pellets to be of inferior quality owing to highly variable, poor composition, excessive moisture content, high temperature of the product, presence of too much meal, and softness.

B. Pellet Quality

1. Factors Affecting Pellet Quality

Different components of cassava affecting its pelletability are starch, fiber, fat, and impurities.[9] Starch gelatinizes when heated in the presence of water and acts as a binder to produce strong pellets. Fibers are difficult to compress but sufficiently fine strands in the pellet impart toughness to the product. Fat merely acts as a lubricant resulting in easy pressing and therefore, low power consumption. The presence of fibers and absence of fats suggest that cassava is a relatively difficult material to pelletize. The presence of sand and other gritty impurities affect not only the quality of pellets but also the life of the dies and rolls.

Conditions of the material such as moisture content, particle size, and temperature before pressing also effect the pelletability. Excess moisture prevents compression and results in what is called "choking". As volume is reduced by pelleting, excess water in the interstices prevents further compression. The choke-up point is reached when the moisture content is around 18%. For the pelletization of cassava, the recommended moisture content lies between 16 and 18%.[9] When fine and medium size particles are mixed, the best quality pellets are obtained. For high-starch materials it has been suggested that the temperature must reach 180°F for proper pelletizing.[9] In the cassava pelleting plants the increase in temperature is achieved purely by the generation of heat due to friction with cassava and hence, results in high wear rate of the die. This could be avoided by spraying the material with steam instead of adding water for moisture content adjustment. The pressure of the steam can be so selected as to give the required temperature and moisture content at the same time.[13]

Die thickness (pelletizing length), die opening diameter, speed of rotation of the die, and the rate of feeding have to be of an optimum combination for production of good quality pellets at low operational cost.[9]

Large chips are not a good starting material for pressing.[13] Field studies carried out in a local pellet mill have indicated that reasonably good pellets could be produced from thin chips, slices, and strips even without any steam injection.[14] Experiments have shown that a quantity of 0.152 kg of steam per kilogram of input material gave dense, glossy, and hard pellets.[14] In this study steam temperature was kept at 120°C for 2 to 3 min exposure to preheat the chips to 70°C before pressing. Pellets produced from chips with a moisture content higher than 17% are of inferior quality. A moisture content of about 14% for the chips before steam admission yields excellent pellets.

Cassava strips having a bulk density of 0.288 g/cc, pressing after addition of steam produced pellets with a bulk density 0.808 g/cc.[14] The savings in space, transportation cost, storage, and packaging cost from the reduction in volume are obvious from these values.

2. Tests for Pellet Quality

Tests with regard to starch, fibers, moisture content, and extraneous matter are the same as for the chips. Two tests have been suggested to determine the hardness and strength of the pellets or to measure the resistance of the material to breakage by impact when handled, dropped, or submitted to heavy burdens. One test known as the "shatter test" can be used for the measurement of resistance to shatter by impact, as during ship loading. Another test, the "Cochrane test" can be used to measure the resistance to abrasion.[9]

In the shatter test a representative sample of known particle size is prepared and dropped from a box to a steel plate. The amount of breakage is measured by screening the product. The percentage of materials remaining on a set of sieves are recorded as the shatter indexes.

The Cochrane test consists of the rotation of sample in the steel drum with angle lifting plates welded inside the drum. The drum is rotated at a constant speed for a set number of rotations and the abraded sample withdrawn and screened on a sieve. The abrasion index is given by the amount of material remaining on the sieve. Cassava pellets are desired to have such "toughness" as to withstand moderately rough handling without excessive breakage.

III. CASSAVA DRYING

Fresh cassava roots cannot be stored under ambient conditions for more than a few days after harvest without spoilage. To overcome this difficulty in the marketing and utilization of cassava and to avoid heavy postharvest losses, the roots need to be processed into some form of dried product with longer storage life. The simplest and the common mode of processing cassava is conversion of the tubers into sundried chips.

A. Sundrying

The sliced tubers are usually dried in the open air under sunlight by spreading in a single layer on cemented floor, bamboo mat, rock surface, or sometimes even on bare earth (Figure 4).[15] Chips dry better on rocks, are white in color, and take less time to dry. Depending upon the weather conditions, it takes 2 to 5 days to dry the cassava chips. Contamination by airborne dust, dirt, and debris cannot be entirely avoided during sundrying especially on windy days.

The chips should be turned periodically during the drying period, until the moisture content reaches 13 to 15%. The chips are considered dry when they are easily broken but too hard to be crumbled by hand. The size of the chips also has an affect on the quality. Thick slices may appear dry on the surface when their internal moisture content is still high.

The recovery rate of chips from roots is expected to be about 38 to 40%. However, it

FIGURE 4. Traditional sundrying of cassava chips in an Indian village.

will depend on the moisture content of fresh roots. To produce white chips of superior quality it is recommended that the roots be trimmed, peeled, and washed in a manner similar to the processing of cassava starch. Otherwise the chips are brown in color and have a high content of fiber, sand, and foreign material.[9,10] The washing of cassava roots before the chipping process would reduce the bacterial content of chips and pellets during the storage.[10]

The common practice in Malaysia and Thailand is to produce the chips during 07.00 and 10.00 hr in the morning. The fresh chips are then spread on the drying floor with shovels and turned every 2 hr to ensure uniform drying. At night and during rains the chips are heaped into mounds and covered. In Malaysia, small tractors fitted with wooden boards are used for the distribution and collection of the chips. In Indonesia, peeled roots are dried by various methods, hung from poles or lines on the farm, or spread out on a rooftop or woven bamboo mats on the ground.

Since sundrying is entirely weather dependent, duration of drying and quality of chips vary widely. The time taken to dry the fresh cassava chips to a moisture content of about 15% is 1.5 days in Malaysia, 3 days in Thailand, and more than 1 week in Indonesia.[9] This is mainly due to the difference in size of the chips produced and the rate of loading. Mathot estimated the overall efficiency of the present sundrying methods to be 13.5% in Malaysia and 11% in Thailand.[9]

B. Tray Drying

Drying on perforated trays is more advantageous in regions with appreciable wind speeds.

Also less manpower is required as chips do not have to be turned and respread to ensure uniform heating.

1. Drying on Shelf Tray

At the Asian Institute of Technology in Bangkok, a three-tier shelf-tray dryer was developed and studied as an alternative to concrete floor drying.[16] The same level of final moisture content (13%) could be attained after 12, 13, and 14 hr when cassava strips were dried on a simple cement floor, the shelf-dryer, and a black-painted cement floor. Floor temperatures of plain cement floor, blacktopped cement floor, and shelf-dryer (upper shelf) were observed to be 33.5, 35, and 30.2°C respectively. Superiority of shelf-dryer was attributed to better circulation of the ambient air through the chips.

Experiments conducted on the influence of shape and size on the rate of drying showed that strips and slices required less drying time than cubes and conventional chips, but the difference was not substantial except in the case of drying on a plain cement floor.

2. Drying on Inclined Mesh Tray

Experiments at CIAT, Colombia, showed that the improved circulation of air permitted higher chip loading rates than on concrete.[17] In this method, the chips are spread on wood-framed trays with a base of 1 in. chicken wire topped with fine plastic mosquito netting. The trays are then placed at an angle of 25 to 30°, which just prevents the chips from sliding down, being supported on a bamboo frame of posts and rails facing the direction of the prevailing wind. Before a rain or at night, the trays can be stacked horizontally, with a canvas or a corrugated iron sheet covering the top tray. Average drying time was observed to be 6 hr and 11 hr at loading rates of 5 and 10 kg fresh chips per square meter. Inclined drying was faster when compared with methods of drying on plain concrete, black concrete, and horizontal trays placed 30 cm above the ground.

Best suggested optimum loading rates on concrete to be 5 to 7 kg/m² and on inclined trays to be 10 to 16 kg/m² depending on the wind speed.[18] Best also observed that if the moisture content is reduced to 50% on the first day, whether dried on concrete floor or on an inclined mesh tray, the cassava would not deteriorate for 3 days after chipping. This may be of considerable importance when a large amount of chips have to be processed during peak seasons of harvest.

3. Drying on Vertical Tray

To take the maximum advantage of the drying power of the wind, trays should be held vertically. Investigations by Roa showed that vertical trays containing chips sandwiched in between two parallel wire panels and positioned against the most predominant direction of the wind, were more efficient than horizontally elevated mesh trays and the conventional system of drying cassava on a concrete floor.[19] Making use of a computer model to interpret his experimental results, he predicted that the maximum permissible chip densities to complete drying within 3 days without deterioration are 5 to 13 kg/m² on a plain concrete floor, 20 to 30 kg/m² for horizontally elevated trays, and 30 to 40 kg/m² for vertically held trays. When covered by a roof, drying rates are not reduced thus, drying may continue overnight or during rains. In spite of the advantages of vertical trays, the additional labor involved in their loading and unloading was considered likely to preclude their adaptation under large-scale commercial processing.

C. Solar Drying

Two types of natural convection solar dryers were tested by Roa for drying cassava chips.[19] One was a solar cabinet consisting of a collector that was made of a series of wire screens painted with a lamp-black base enamel and covered by two sheets of window glass. A vertical wood frame prevented lateral heat loss and provided support to the collector and glass sheets. The product was placed just below the collector on a perforated base.

The second solar drier tested consisted of the same glass sheet arrangement with a collector made of a black metal sheet, with a fin system soldered on its back. The chips were placed in a perforated tray such that the collector fins rested on the particles.

However, the performances of the solar cabinet and the solar dryer tested with regard to drying the chips did not result in appreciable advantages over natural drying, although high radiation levels were recorded during the tests. Roa[19] implied that attempts to collect the solar energy restricted the air circulation which had more energy available for drying a high-moisture product such as cassava chips, particularly during the early periods of the dehydration process.

Attempts have also been made to achieve satisfactory solar drying of cassava chips by restoring to forced convection. In this context a through circulation solar-heated dryer constructed at CIAT, Colombia, has successfully been used to dry loadings of 125 kg/m² in 45 hr.[20] The drying bin constructed of brick, had a 1 m × 2 m concrete floor. The bin, 1.7 m in height, had a perforated floor of a galvanized steel sheet supported by wooden beams at a height of 60 cm above the floor. The bin was also protected from air. A solar collector with an area of 10 m² was constructed on a concrete base and concrete blocks placed over a 20 cm thick bed of fine stones, both painted black. A polyethylene cover was laid over the blocks. A centrifugal fan driven by a 1.5 hp motor and having an airflow capacity of 38 m³/min had been installed to draw air from the collector into the bin. Results of the trials, carried out to determine the optimum time of starting drying, clearly indicated the advantage of commencing drying in the later hours of the afternoon in terms of drying time and the rate of loss of moisture. This is so because during night hours, even when the relative humidity is high, moisture can be removed from the fresh cassava chips and then the hottest hours are available to terminate the process.

Shrinking of the drying bed was quite noticeable. For a loading rate of 125 kg/m² the thickness of the layer of cassava chips decreased from about 25 to 18 cm. The pressure drop through the drying bed, an important factor for selecting an appropriate blower, never exceeded 8.4 mm (0.33 in.) water gauge. It was in fact, observed that the pressure drop decreased during drying as the resistance to airflow was reduced, presumably because cassava chips lost their tendency to adhere to each other.

The solar collector in this case gave average increases of 1 to 3°C in air temperature corresponding to reductions of 4.4 to 9.6% in relative humidity while the airflow rate used was 18 m³/min.[2]

In Indonesia, dried cassava, known as gaplek, is traditionally produced in whole form or halves of cassava roots without any skin. In the hang-drying method, cassava is cut into 4 halves, hung on a rope, and dried for 2 to 3 days. The traditional method of drying on bamboo matwork takes 7 to 10 days. Experiments on the use of solar dryers in the gaplek-producing process have been conducted in Indonesia using two types of solar dryer with a capacity of 200 kg cassava.[21] These solar dryers were 3 × 1 × 2.35 m in dimension and 0.5 m above the ground. Each dryer had 7 racks with 30 cm interspacing and two trays in a rack. Walls of type-1 were made from transparent polyethyene, except the south wall which was made from black polyethylene. However, type-2 had a separate flat plate collector attached to the drying chamber trays and had walls of transparent polyethene. The solar dryer of type-2 could produce white gaplek of 12 to 13% moisture content after 4 days of drying which was faster than the type-1 as well as sundrying. Highest temperatures attained by the solar dryers type-2 and -1 and sundrying at 12-noon were 41.8, 39.7, and 34.3°C. The moisture content of gaplek in cube form was lower than that of halved gaplek. Between type-1 and sundrying there was no significant difference. In order to achieve uniform drying and to arrest mold growth in the middle racks, use of 3 to 4 trays were recommended.

D. Artificial Drying

It appears that the greatest problem in natural drying, whether on concrete or in trays, lies in the reduction of the moisture content from around 35% to a safe storage value below 14%. Although this range represents only about 25% of the total water content of cassava, its removal can occupy up to half the drying time.[17] This problem can be overcome by replacing, or at least combining, natural drying with the use of either solar heated air dryers or artificial dryers to give greater operational flexibility, with the latter eliminating the dependence on weather as well.

Heat and airflow characteristics of cassava were first studied by Ghosh,[22] with the help of a laboratory artificial dryer set at three levels of moisture content, using two different air velocities, and keeping the maximum drying air temperature at 80°C. Temperature changes were measured during drying at the top, middle, and bottom layer of a 45 cm (18 in.) deep bed. It was found that rise in temperature of the chipped cassava only began when the moisture content had been reduced to around 35% which took about 6 hr of drying. The drying rate and temperature rise at the top layers were found to be considerably lower than that at the bottom. Higher inlet air velocities for a given drying temperature increased the drying rate to an appreciable degree.

Chirife and Cachero[23] studied the optimum drying parameters for cassava in a laboratory through-circulation dryer. Experiments were done by cutting fresh cassava into uniform rectangular slabs of 6 to 7 × 2 × 0.3 cm size. Variables studied were bed depth (2 to 12 cm), air velocity (0.5 to 1.4 m/sec corresponding to an airflow rate of 2300 to 5200 kg/hr m^2), air temperature (55 to 100°C), and also static pressure drops of air passing through beds of wet and dried slices.

Results showed that the output increased with the increase in bed depth, air velocity, and air temperature. However, scorching of cassava chips occurred at 85°C and above. Temperature of the chip at the surface was found to increase continuously right from the commencement of the drying process, indicating that there was no constant-rate drying period. Drying occurred wholly within the falling rate period. In fact two falling rate periods were observed. The optimum conditions recommended for thorough circulation drying of cassava chips are a bed depth of 12 cm, air mass flow rate of 4500 kg/hr m^2, and air temperature less than 84°C.

As the internal water movement was assessed to be the controlling mechanism from the beginning of the process of cassava drying, Chirife analyzed the experimental results in relation to various mechanisms of internal flow likely to operate and confirmed that moisture migrates within cassava root by a process of liquid diffusion,[24] rather than by capillary flow, viscous flow, gaseous diffusion, thermal diffusion, or other mechanisms.

Webb and Gill,[25] investigated laboratory scale artificial drying of cassava chips of 4.5 × 0.35 cm size using air temperatures of 55, 66, and 77°C, and bed depths of 5, 8, and 10 cm, and found that there was no initial constant rate phase but two distinct falling rate phases governed the moisture diffusion rate within the chips. Since the rate of mass transfer by diffusion generally increases with temperature, these workers advocated the use of the highest possible product temperature consistent with the avoidance of scorching and other physio-chemical changes in the product. They employed a two-stage drying process in which the air temperature was 94°C in the critical phase, but was later reduced to 66°C to avoid scorching.

A medium scale and simple drying plant is reported to have been recently developed,[26] using two bins of 4 m × 7.5 m size, fitted with perforated floors which can be constructed on site using concrete block walls with timber floors and doors, or from other construction materials. With a bed depth of 1 m and a heat input of 12,600 M$_j$/hr, 5 t of dried chips can be produced per day using the two bins alternately.[26]

Efforts to develop artificial drying of cassava chips are understandable if one considers the advantage of this method over sundrying. Among the significant advantages are

1. Reduction in requirements of space, labor, and time.
2. Riddance from weather-dependent scheduling of the operations such as harvesting and processing of cassava.
3. Better quality control with regard to final moisture content and microbial spoilage.

The investment in artificial heat drying facilities together with fuel, power, and maintenance costs should be weighed against the investment in land and cement drying yard together with the cost of labor.[9]

There are a large number of dryers that can be used for cassava products. Static-bed dryers, which include bin, tray, and through-circulation dryers, are commonly used for drying grain. Within existing constraints, the most suitable artificial dryers for cassava chips are through-circulation batch dryers.[17] These dryers, though low in heat efficiency, are simple and inexpensive. Moving bed dryers are mostly suitable for free-flowing materials such as grains and their application for drying a particular type of cassava chip must always be experimentally ascertained before hand. Fluidized-bed dryers may be found suitable for cassava chips of smaller particle size or for partially dried ground cassava which can be easily crumbled into loose particles.

At present, methods of cassava drying remain dependent on weather. Grace suggests that in rainy regions, where continuous sundrying of cassava is not possible, some form of artificial drying is required.[3] Mechanical drying systems may produce a higher quality product but these are also associated with higher costs. In spite of higher thermal efficiency these may still be too expensive for production of cassava chips. However, the potential of artificial cassava drying systems may still lie in its optimal combination with natural drying.

E. Equilibrium Moisture Content of Cassava

The equilibrium moisture content (EMC) is directly related to the drying and storage of any farm crop. It is used to determine the final moisture content that an agricultural product will attain under a given set of temperature and relative humidity conditions.

Published information on the moisture content relative humidity equilibria of dried cassava products is scant and inconsistent. Relationship between air temperature and relative humidity corresponding to equilibrium moisture contents of 10, 14, and 25% (wet basis) in cassava has been given by Webb and Gill,[25] but the experimental procedure used was not described. For example, one interpolation of this curve, shows that an equilibrium moisture content of 14% corresponds to 70% relative humidity at 30°C.

Investigations carried out at TPI (present TDRI), London, on EMC of cassava chips used the static method of saturated solutions of inorganic compounds which can maintain an atmosphere of known relative humidity in a thermostatically controlled closed surrounding. When the cassava chips reached a constant weight, which took about 3 weeks time, its moisture content was determined by heating to constant weight at 110°C. The results are given in Table 2.

When a solid is exposed to a continual supply of air at constant temperature and humidity, the solid will either loose moisture by evaporation (e.g., as in drying) or gain moisture from the air (e.g., as in storage) until its moisture context is in equilibrium with the surrounding.

Knowledge of EMC is not only essential in the application of thin layer and deep bed drying equations but can also be used to predict the natural drying and open storage behavior of a crop commodity. For example, from EMC values of cassava chips computed for average ambient temperature and relative humidity using 10 years (1973 to 82) climatological data of Trivandrum in India, it could be found that at no time during the year the EMC values of cassava chips go below 14.6% (Figure 5) rendering unheated air drying as well as open storage conditions unfavorable.[28]

Table 2
EQUILIBRIUM MOISTURE CONTENT
OF CASSAVA[27]

Temperature (°C)	Relative humidity (%)	Equilibrium moisture content (% wet basis)
30	28.0	9.4
	36.8	10.3
	44.6	10.8
	54.2	12.2
	63.4	13.9
	74.9	16.1
	84.5	18.2
45	21.5	7.2
	30.2	8.9
	44.8	9.7
	54.3	10.9
	73.0	15.2
	80.0	16.6

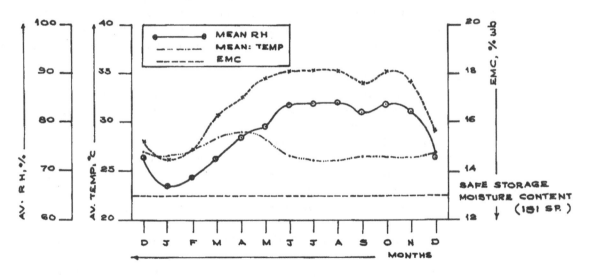

FIGURE 5.   Variation of EMC of cassava chips with temperature and relative humidity throughout the year at Trivandrum.

REFERENCES

1. **Anon,** *CIAT Highlights in 1978,* Centro Internacional de Agricultura Tropical, Cali, Colombia, 1979, 27.
2. **Both, R. G.,** Post-harvest handling and storage. IV. Root crops, fruits and vegetables, *Agribusiness Worldwide,* Sept.—Oct., 12, 1984.
3. **Grace, M. R.,** Cassava Processing, Plant production and Protection Ser. No. 3, Food and Agriculture Organization, Rome, 1977, 155.
4. **McFarlane, J. A.,** Cassava storage. II. Storage of dried cassava products, *Trop. Sci.,* 24, 205, 1982.
5. **Lancaster, P. A., Ingram, J. S., Lim, M. Y., and Loursey, D. G.,** Traditional cassava based foods: a survey of processing techniques, *Econ. Bot.,* 36, 12, 1982.
6. **Anon.,** *Wealth of India — Raw Materials,* Vol. 6, Coun. Sci. Ind. Res., New Delhi, 1962, 330.

7. **Nanda, S. K.,** *Hand-Operated Cassava Chipping Machine,* Publication No. 5, Central Tuber Crops Research Institute, Trivandrum, India, 1983, 6.
8. **Nanda, S. K.,** Field evaluation of hand-operated cassava chipping machine, in *Proc. Natl. Symp. Prod. Utilization Trop. Tuber Crops,* Indian Society for Root Crops, Trivandrum, India, 1985, 38.
9. **Manurung, F.,** Technology of cassava chips and pellets in Indonesia, Malaysia, and Thailand, in *Proc. Interdisciplinary Workshop Cassava Proc. Storage,* Araullo, E. V., Nestel, B., and Campbell, M., Eds., Pattaya, Thailand, 1974, 89.
10. **Thanh, N. C.,** Technology of cassava chips and pellets processing in Thailand, in *Proc. Interdisciplinary Workshop Cassava Proc. Storage,* Araullo, E. V., Nestel, B., and Campbell, M., Eds., Pattaya, Thailand, 1974, 113.
11. **Odigboh, E. V.,** Cassava chips processing and drying: a cassava chipping machine, in *Proc. Workshop Small-Scale Proc. Storage Trop. Root Crops,* Plucknett, D. K., Ed., Westview Press, Colo., 1979, 243.
12. **Orias, R. R. and Cruz, R. O.,** Root crop processing: assessment for village level adaptability, in *The Radix,* Root Crop Research and Training Centre, Leyte, Philippines, 2, 3, 1981.
13. **Mathot, P. J.,** Production and export control in Thailand and the marketing in Europe of tapioca pellets, in *Proc. Interdisciplinary Workshop Cassava Proc. Storage,* Araullo, E. V., Nestel, B., and Campbell, M., Eds., Pattaya, Thailand, 1974, 27.
14. **Thanh, N. C., Muttamara, S., and Lohani, B. N.,** Technological improvement of Thai tapioca pellets, *Thai J. Agric. Sci.,* 11, 1978, 75.
15. **Nanda, S. K.,** Post-harvest practices of tuber crops, *Indian Farming,* 33, 64, 1984.
16. **Thanh, N. C. and Lohani, B. N.,** Cassava chipping and drying in Thailand, in *Proc. Workshop Cassava Harvesting Storage,* Weber, E. S., Cock, J. H., and Chouinard, A., Eds., International Dev. Res. Centre, Ottawa, Canada, 1978, 21.
17. **Best, R.,** Cassava Processing for animal feed, in *Proc. Workshop Cassava Harvesting Storage,* Weber, E. S., Cock, J. H., and Chouinard, A., Eds., International Dev. Res. Centre, Ottawa, Canada, 1978, 12.
18. **Best, R.,** Cassava Drying, Series O5EC-4, Centro Internacional de Agricultura Tropical, Cali, Colombia, 1979, 24.
19. **Roa, G.,** Natural Drying of Cassava, Ph.D. thesis, Michigan State University, East Lansing, 1974, 234.
20. **Best, R., Alonso, L., and Velez, C.,** The development of a through-circulation solar-heated air drier for cassava chips, in *Proc. Symp. Int. Soc. Trop. Root Crops,* 6th ed., Shideler, F. S. and Rincon, H., Eds., International Potato Centre CIP, Lima, Peru, 1984, 373.
21. **Radiyati, T., Winarto, A., Gardjito, M., Muljohardjo, M., Suksmadji, B., and Sardjono,** Use of solar dryer in Gaplek producing process, in *Proc. Regional Asia Pacific Workshop Applications Solar Energy Agric. Post Harvest Technol.,* Harahap, F. and Abdurrechim, H., Eds., UNESCO, Regional Office for Science and Technology, Jakarta, 1981, 177.
22. **Ghosh, B. N.,** Heat and airflow characteristics in drying crops, in *Proc. Int. Symp. Trop. Root Crops,* Tai, E. A., Charles, W. B., Haynes, P. H., Iton, E. F., and Leslie, K. A., Eds., University of the West Indies, St. Augustine, Trinidad, 1969, VI-1.
23. **Chirife, J. and Cachero, R. A.,** Through-circulation drying of tapioca root, *J. Food Sci.,* 35, 364, 1970.
24. **Chirife, J.,** Diffesional process in the drying of tapioca root, *J. Food Sci.,* 36, 327, 1971.
25. **Webb, B. H. and Gill, K. S.,** Artificial heat drying of tapioca chips, *Malays. Agric. Res.,* 3, 67, 1974.
26. **Coursey, D. G., Marriott, J. A., Mc Farlane, J. A., and Trim, D. S.,** Improvements in the field handling, chipping and drying of cassava, *J. Root Crops,* 8, 1, 1982.
27. **Anon.,** *Relationship between Atmospheric Humidity Temperature and the Equilibrium Water Content of Cassava Chips,* L1, Trop. Prod. Inst., London, 1965, 4.
28. **Anon.,** *Two Decades of Research,* Central Tuber Crops Research Institute, Trivandrum, India, 1983, 154.

Chapter 7

CASSAVA IN ANIMAL FEED

During recent years there is a remarkable increase in livestock production in the third world. Because of this, the requirement for animal feed materials is also increasing. The spiraling price hike of cereals has necessitated the search for alternate sources of energy for animals. Being a cheap carbohydrate source capable of supplying adequate calories, cassava tubers offer great potential as an animal feed. However, due to certain limitations such as its low content of proteins, vitamins, and some minerals, and lack of sulfur-containing amino acids such as methionine, it is often rated as inferior to maize or wheat. Research on the use of cassava as an animal feed has been carried out for almost a century. These studies definitely show the importance of cassava in animal nutrition. However, the presence of hydrogen cyanide (HCN) has to be properly recognized and its level has to be minimized by various processing techniques. Proper formulation of the diet is equally important to make the feed nutritionally balanced, since the animal performance is highly dependent on it. Cassava leaf meal is a highly nutritious protein-rich ingredient which offers a vast scope for inclusion in root meal diets. However, the leaf meal has to be properly detoxified by drying prior to its inclusion in compounded feeds. Cassava roots and leaves have been found to contribute substantially to the energy requirements of poultry, swine, and ruminants. Cassava diets have often been found to have low palatability due to the powdery nature of the root flour. However, proper supplementation with molasses and/or fat has been found to reduce dustiness and improve taste.

I. CASSAVA AS FEED SOURCE FOR POULTRY

The incorporation of cassava feeds for chick production started with the experiments of Tabayoyon,[1] who extracted a product from cassava starch and fed this to chickens with other ingredients such as rice bran, corn meal, and shirmp meal. When this material was incorporated at levels of 30 and 60% in poultry diets, the performance of birds was found to be satisfactory although slight reduction in feed consumption and weight gain were noticed at higher level. Contrasting reports on the metabolizable energy (ME) of cassava root meal are available. Olson et al.[2] found calorific values of 3.44 kcal/g of dry flour while Maust et al.[3] reported values of 4.31 kcal/g. Muller et al.[4] found values of 3.65 kcal/g while Hutagalung reported values of 3.23 kcal/g. The age of tubers and method of processing also influence to a great extent their ME values.[6] The difference in the product as well as varieties may be responsible for the variation in ME values.

The relative digestibility of cassava root flour fractions by poultry was investigated by Vogt and he found that crude protein digestibility was 75%, crude fat 70%, crude fiber 55%, and nitrogen-free extract 99%.[7] Studies on the maximum levels of cassava incoporation in broiler diets were made by a number of workers. McMillan and Dudley[8] found no deleterious effect on chickens fed 20 and 40% cassava root meal (CRM). Lower levels of up to 10% were recommended by Klein and Barlowen and Wegner.[9,10] Vogt and Penner found that productivity of broilers was normal with 10% CRM but weight gain and feed efficiency decreased with 20 and 30% CRM.[11] Broilers should be fed a low level of cassava, viz., 10% up to the 4th week.[7,12] Manjarrez et al.[13] prepared "cassava rice" with 40% cassava and 60% rice and this was incorporated at levels of 50 and 100% in broiler diets. Montilla et al.[14] found that feed efficiency decreased as the percentage of inclusion of cassava increased, probably as a result of the powdery characteristic of the CRM. Hence, the effect of 5% animal fat and 5% sugar cane molasses to CRM was studied by Montilla et al.[15] The deterioration in feed efficiency could be annulled by this treatment.

The excessively fine nature of cassava root meal could be reduced by pelletization which has also been found to improve feed intake.[16] In nutritionally balanced pelleted cassava feed, up to 58% replacement with cassava was reported to give satisfactory performance of broilers.[17] Phuah and Hutagalung[18] found that the percentage of carcass yield and carcass protein were significantly higher and production of fat was lower with 20% cassava root flour than with 40%. They concluded that there were no adverse effects if the diet was properly supplemented with methionine, lysine, and palm oil.[18] Salt-treated cassava chips produced diarrhea and mortality of broilers. Many studies show that cassava rations promote growth of broilers and optimum growth occurs at a protein level of 19 to 20%.

It has been reported that laying birds can tolerate up to 40% cassava meal. However egg production was found to be reduced if the diets are not well-balanced.[19] In contrast to this, an increase of 19.2% in egg production was reported by Falanghe when wheat bran was replaced by cassava meal and rice bran.[20] Optimal level of cassava inclusion in layer diets has been reported by several workers as 20%.[7,21,22] However, the low protein content of cassava meal has to be compensated by supplementing with high protein feed ingredients. Replacement of 50% ragi flour with cassava spent pulp has been reported to increase egg production by 12%.[23] Replacement of up to 60% of a corn diet with cassava has been reported not to affect feed efficiency and egg production by Hamid and Jalaludin.[24] Higher levels of cassava inclusion necessitate proper supplementation with fish meal.[25] In a 40-week study, Eshiett and Ademosun[26] found that varying levels of cassava up to 40% do not affect the feed conversion, egg production, egg weight, etc. from feeding trials conducted on starter and grower chicks. Ademosun and Eshiett concluded that their diets should not contain more than 15 and 30% CRM respectively.[27]

Hamid and Jalaludin[24] found that egg yolks became progressively whiter as the percentage of incorporation of cassava was increased in the ration. Similar results have been reported by Enriquez and Ross.[28] The principal factors contributing to the whitening of egg yolk are the low fat content of cassava flour meal and the low content of pigmenting xanthophyll in the cassava diet. However, when yellow cassava was used as a feed source, the lack of yolk pigmentation could be solved to some extent.[29] Another approach in this line is the incorporation of cassava leaf meal in tuber meal diet. Agudu compared the efficacy of cassava leaf meal, madras thorn leaf meal, a synthetic xanthophyll, and two types of corn in pigmenting egg yolks.[30] He found that cassava leaf meal had a high total and pigmenting xanthophyll content (605 and 508 mg/kg respectively), compared to madras thorn leaf meal and even commercial xanthophyll.

The low fat content of cassava meal also has an effect on egg weight. Hens fed purified diets lay smaller eggs than those fed practical ration. This defect could be remedied only by the addition of soybean or fish meal protein.[31] Corn oil was reported to increase egg weight in cassava rations.[32] As compared with 60.8% linoleic acid in corn fat,[33] cassava fat contains only 14.6% linoleic acid.[34] Moreover yellow corn contained 42 g fat/kg DM while cassava contained 5 g fat/kg DM. Thus, it is quite possible that linoleic acid deficiency may be created in cassava rations. This can be compensated by the proper addition of fat.

Dietary methionine has been reported to be another factor contributing to egg weight. Oyenuga found that while maize contained 194 mg/16 g N of methionine, cassava contained only 63 mg/16 g N.[35] This can lead to a deficiency of methionine in cassava rations which can adversely affect the egg size. Protein and amino acids are the major factors contributing to egg size and hence, proper care has to be taken to balance these nutrients in cassava rations to obtain optimum performance of layers fed this diet.

Shell thickness is another factor contributing to egg quality. Vitamin C has been reported in some cases to improve shell thickness.[36,37] El-Boushy et al.[38] also found that vitamin C addition improved shell thickness and shell structure. However, cassava is fairly rich in vitamin C (35.0 mg/100 g as compared to 11.4 mg/100 g DM in maize) and hence, this need not be a limiting factor in cassava rations.[35] Enriquez and Ross found that shell thickness was increased by increasing the percentage inclusion of cassava.[28]

Hoyos and Santos[39] observed that the feed conversion efficiency of a cassava diet could be considerably improved by the addition of vegetable oil. Another factor contributing to increased feed intake was pelletization of the mashed cassava meal.[40] Jalaludin and Leong[25] found that higher levels (75%) of cassava increased the feed intake but feed efficiency was less. However, a high percentage (up to 13%) of fish meal has to be incorporated into the diet. Khajarern et al.[41] found that cassava could economically substitute corn in broiler rations only when the price of cassava was 60% that of corn. Phalaraksh et al.[42] reported that under heat stress conditions, cassava-fed pullets were more tolerent to stress and showed a lower loss (42%) in terms of mortality than those fed corn, broken rice, etc. No adverse effect on egg weight, albumen height, shell thickness, etc. could be noticed in birds fed 60% CM during the growth phase and 50% during the laying period. However, egg production on the hen/day basis was slightly lower than the corn control.

Cassava leaf meal has been reported as a high protein ingredient which can be included in poultry rations. Miranda et al.[43] compared the effect of alfalfa, cassava leaf, tropical kudzu hay, etc. at 5% level and found no difference between the leaves. Ross and Enriquez[44] incorporated up to 20% cassava leaf meal in growing chick rations. The leaf meal was prepared by drying the leaves at 50°C. Methionine and energy appear to be the two prime factors limiting the incorporation of cassava leaf meal in chick feed. They found that when 0.15% sodium thiosulfate was added to 20% cassava leaf meal significant growth improvement could be produced. Siriwardene and Ranaweera,[45] prepared a composite flour from leaf blades, stalks, and young stems of cassava, and this flour having 22.3% protein was fed to chicks up to a 10% level. They could not obtain any significant difference in body weight gain and feed efficiency.[45] They further found that cassava foliage meal could be incorporated at high levels in pelleted broiler rations.

A. Additives Added to Cassava Based Diets for Poultry

Much of the studies on the type of additives added to cassava diets have been restricted to methionine since this is the primary limiting factor in cassava root meal rations. Moreover, methionine helps in the detoxification of cyanide by providing labile sulfur and also another sulfur containing amino acids such as cystine and cysteine. Several studies have shown that the protein from cassava leaves and roots contain insufficient quantities of sulfur containing amino acids.[44,46,47] Since groundnut cake or soybean meal which are usually added to cassava rations as protein supplements are also deficient in methionine, supplementation of such feeds with methionine would be desirable for optimal performance of poultry. Initial reports on the supplementation of cassava root meal with methionine show that consistent improvement in body weight resulted from addition of 0.2% DL-methionine.[48] In cassava leaf meal (20%), 0.3% methionine was found to be essential for optimum performance. Increased urinary thiocyanate values were obtained from animals fed methionine.[49,50] Job[51] found that tiosulfate was a better source than methionine for cyanide detoxification. However, the cost of supplementation has to be taken into account while computing the net again in growth improvement. Batatunde et al.[52] observed that under conditions in Nigeria, a high protein diet unsupplemented with methionine was more economical to feed broilers than methionine supplemented with low protein diet.

Additives other than methionine added to cassava-based diets include fats and oils, protlins, nonprotein nitrogen sources, minerals, vitamins, etc. Growth depression of animals fed cassava diets was found to be corrected by the addition of fats and oils.[53-55] The effect of palm oil and stearin in feed gain and efficiency of poultry has been studied by several workers.[56-59] They found that in terms of digestibility palm oil was superior to fats of animal origin in cassava rations. Up to 8% palm oil supplementation in cassava diet has been successfully tried in poultry rations. The main effect of vegetable oil was the improvement of feed conversion.[60]

Since cassava root meal (CRM) is low in protein, when it is used in large proportions, much of the protein should come from other high protein sources. Soybean meal has been tried in some cases, however, animal proteins like fish meal are performing better.[61,62] Cassava leaf protein is an excellent protein supplement since its digestibility is also high. Young leaves had a true digestibility of 80% for the protein while older leaves had only 67% digestibility.[63] The cassava leaf protein digestibility in rats has been reported to be 70 to 80% and its biological value was 44 to 57%. Another protein source is fish meal which has to be added at the rate of 10% to a 50% CRM ration.[64] Significant improvement in growth could be obtained by supplementation of cassava root meal with groundnut flour, skim milk powder, or a mixture of the two.[65] Olson et al.[66] studied the effect of leucine supplementation to rations containing 45% cassava meal in chicks and found that the weight gain was better when leucine was supplemented along with methionine. Muller et al.[64] also found that addition of 15% poultry excreta to cassava based diets produced favorable responses.

Supplementation of cassava diets with various minerals has been attempted by many workers. Although the calcium and phosphorus contents of cassava roots are relatively high the contents of other minerals are low, especially of copper, iron, and zinc.[3,67-69] The excess addition of soybean meal to cassava meal to elevate the protein content of the composite feed can lead to impaired absorption of minerals like calcium and zinc due to formation of insoluble phytates with the phytic acid in soybean meal.[70-72] The presence of calcium in excess in cassava root meal and fish meal can also hamper the availability of zinc by altering the intestinal pH.[73] Hence, in such diets it is always desirable to add zinc. With supplementation of iodine, the diet can reduce the chances of cyanide toxicity and can prevent the thyroidal depletion of iodide stores. Hence, diets containing a large proportion of cassava have to be properly supplemented with iodine. The extent of supplementation is important since excess iodine can also lead to cessation of egg production, delayed sexual maturity, diarrhea, etc.[74-46] Supplementation of the diet with iron also has to be done with caution, since excess iron can impair the absorption of P by forming insoluble phosphates.[77,78]

In addition to the low protein and fat content of cassava roots, the low vitamin content sometimes necessitates extraneous addition of vitamin premixes in cassava-based diets. De Brochard et al.[79] reported the vitamin content of cassava root meal to be vitamin A 550 IU/kg, vitamin D_3 0.01 IU/kg, thiamine 1.6 mg/kg, and riboflavin 0.8 mg/kg. Ascorbic acid content in fresh roots ranges from 5 to 360 mg/kg.[68,80,81] Fresh cassava leaves contain high levels of ascorbic acid (0.4 to 1.8 g/kg) and carotene.[68] Niacin and vitamin B_{12} are the two vitamins which need to be incorporated in cassava rations. However, Muller et al.[64] found that addition of niacin to cassava-based diet did not improve the growth performance of chicks. This might probably be due to the adequate supply of these through the fish meal provided in the diet. Biotin addition, on the other hand, was found to improve broiler performance fed a cassava-protein diet.[64] Supplementation of a 20 to 30% cassava diet with a mixture of niacin, panthothenate, and choline chloride was not effective in improving feed conversion.[11] Carvalho et al.[82] found that the perosis observed in cassava-fed broilers could be corrected by choline supplementation in the diet.

Another problem usually encountered by feeding high levels of cassava is the formation of pale soft and exudative (PSE) meat. Meat color is primarily decided by the myoglobin content of meat. Hence, if adequate protein supplementation is there, PSE meat should not be a problem with cassava diets. However, vitamin E addition to the diet has been reported to improve meat quality.[83] Low pigmentation of skin, shanks, and fat of broilers fed a cassava diet has been reported by several workers. Egg production and egg yolk pigmentation are also badly affected in cassava feeding.[84,85-87] However, these defects can be rectified by supplementing the diet with synthetic carotenoids.[67,84,86,87] Cassava leaves also were highly effective in pigmenting the meat and egg yolks.[58]

Khajarern and Khajarern[88] found that the limiting factors in maximum replacement and economic feasibility of cassava root for corn were fiber and protein content, the prices of cassava, corn, and protein supplements such as fish meal and soybean meal.

II. CASSAVA AS FEED FOR SWINE

Although cassava is an excellent energy source for swine, its use in compounded swine feeds largely depends on the price and nutritive value of protein supplements. Digestibility of cassava based diets for swine has been reported to be similar to cereal based diets.[89,90] Aumaitre[91] found that cassava improved organic matter digestibility by 4 to 5% compared with wheat or barley based diets. The crude fiber and ash should be strictly limited in cassava-based diets for pig as the crude fiber can interfere with the absorption of zinc. Maust et al.[92] and Hutagalung et al.[67] observed parakeratosis in pigs fed cassava rations for 4 weeks. This has been correlated with the zinc deficiency created by the excessive fiber content of cassava diets.[93] The high crude fiber content can also make the feed bulky thereby decreasing feed consumption and causing irritations of the eyes and respiratory organs.[64] Pelletization of swine feed could improve the digestibility of starch and crude fiber.[64] However, fresh sweet cassava is consumed readily by gestating gilts when it is properly supplemented with protein, minerals, and vitamins. Life cycle feeding of pigs using sweet cassava meal has been tried by Gomez et al.[94] The crude protein level was maintained as 16% for growing (20 to 50 kg), 13% for the finishing (50 to 90 kg), and 13% for the gestating and lactating pigs.[94] Soybean meal was used as the protein supplement and the long-term effect of cassava feeding on the reproductive performance of gilts was analyzed. Gomez[95] found that the gilts fed cassava meal gained less body weight compared to those fed common maize. However, it was observed that these gilts went on gaining weight throughout the lactation period where as the control gilts lost weight during the lactation period.

There was significantly lower number of weaned pigs in the cassava meal fed group. Methionine supplementation was not found to have any effect on the reproductive performance of gilts fed cassava meal.[95] It was found that the amount in the diet required to produce a weaned pig was 45% higher for the cassava feeding system compared with the common corn feeding system. Gomez et al.[96] observed that piglets preferred and consumed greater amounts of 40% cassava meal compared with 20% cassava meal. No significant difference in average daily gain, feed intake, and feed efficiency could be noted in pigs fed different cassava levels. But the pigs fed a control ration had a higher loin eye area than those on the cassava rations. The moisture, crude protein content, and fat content of the meat and liver of cassava-fed pigs did not differ from the control pigs. Hence, this suggested that cassava root meal can be used as a main carbohydrate source in pig rations and up to 60% cassava meal can be well tolerated by the pigs. In fact cassava was found to have a beneficial effect on the quality of pork by Castillo et al.[97] who reported that the carcass quality, vitamin A of plasma and liver, as well as the back fat thickness were all similar for the cassava-fed and control pigs. Maner and Gomez also obtained similar results.[49] Cassava was reported to improve the organic matter digestibility of pigs by a number of workers.[89,91,98] Contrary reports indicating[89] low performance of pigs fed cassava diets are also available where the percentage shrink of the carcasses was slightly lower for the control pigs than the cassava-fed pigs.[99,100] There was also a slight difference in the saturation of fat in the cassava-fed pig.

Experiments by Acuna have shown that cassava can replace up to 60% yellow maize without affecting the performance of pigs.[101] About 11% reduction in body weight gain has been reported by the total replacement of maize by cassava.[102] The effect of replacing barley with cassava meal in the weaning diet of pigs was studied by Arambawela et al.[103] and it was observed that there was a higher incidence of digestive disturbance in the cassava-fed piglets. This has probably arisen from the increased feed intake observed in this group. Asico had also observed that cassava was only 90% as efficient as corn in the weaning diet of piglets.[104] Peanut oil meal was used as a protein supplement in cassava rations for fattening swines by Barbosa et al.[105]

Compared to CRM, sago flour gave only lower growth rates although the cost of a sago ration was cheap.[106] Fullerton studied the replacement value of cassava meal in barley/corn rations of fattened pigs and found that high grade cassava meal was satisfactory with regards to taste and other dietetic properties.[107] Supplementation of cassava meal with molasses was found to improve average weight gain and reduce the amount of protein supplement needed. Moreover, taste and consistency to the ration can be maintained by the addition of molasses.[108] Hansen et al.[109] found that it was not economical to include more than 20 to 30% cassava meal in the rations of bacon-type hogs. Similar results were reported by Hofman.[110] Kitpanit and Bunsiddhi[111] reported that casava + 17% soybean meal had a nutritive value similar to rice brain + broken rice and could therefore very well replace these in swine rations.

Pelletization of a high cassava meal diet had no improved effect on pig performance.[112] Apparent digestibility of cassava by pigs was found to be 36.4% for crude protein, 44.3% for crude fat, 54.5% for crude fiber, and 95.6% for N-free extract.[113] They also found that cassava contained 87.2% TDN and 3760 kcal digestible energy per kilogram of dietary methionine.

The role of supplementary fat and methionine in swine rations was investigated by Zoby et al.[114] The addition of methionine was found to improve the weight gain, daily feed consumption, and feed efficiency. However, this difference was noticeable only after the pigs reached 50 kg. Supplementation with fat increased the fat content of carcasses. However, Job found that methionine, thiosulfate, and elemental sulfur had no effect in promoting feed conversion efficiency, daily weight gain, etc.[51] Yet, Hew and Hutagalung[54] have observed that 0.2% methionine added to 50% cassava-based diet significantly improved the performance of swines. Similar results for added methionine have been reported by Maner and Gomez.[49] Maner also found that the slight depression observed in a .55% cassava meal molasses fat diet could be overcome by adding 0.1 or 0.2% methionine.[50] The fat and molasses reduce dustiness and improve the palatability of the diet, respectively, while methionine supplies labile sulfur for the detoxification of cassava cyanogenic glycosides.

The efficacy of cassava leaf meal as a protein supplement in swine rations was tried by Lee and Hutagalung.[55] They found that 10 to 20% cassava leaf meal reduced the palatability as well as depressed the weight gains and feed conversion. They have attributed the reasons for this growth depression to the dry and loose texture of cassava leaves and the high crude fiber content of the leaves.

Hutagalung et al.[59] studied the effect of various fat sources like palm oil, stearin, lard, and tallow added to a cassava-based diet on the performance and carcass quality of growing and finish pigs. Palm oil or stearin at the rate of 5 to 10% added to a 30% cassava based diet improved the weight gain and feed efficiency of pigs compared with basal diet and the tallow and lard-supplemented cassava diet. Devendra and Hew[115] further studied the effect of graded doses of palm oil on the performance of pigs fed cassava meal. They observed no significant difference by supplementing the diet with 5, 10, 20, 25, and 30% palm oil. Thus, it was concluded that the optimum level of palm oil addition to a growing pig's diet was 5%. Shimada et al.[116] also found that corn oil (3%) added to cassava diets improved the performance of growing pigs. The importance of animal proteins in cassava rations was brought about from the studies of Hew and Hutagalung[117] who found that the growth depression observed in pigs fed a high percentage of cassava meal could be overcome by maintaining a constant level of plant protein and increasing the animal protein levels.

The ability of sulfur containing amino acids to alleviate the depressive effect of a high cassava diet has been brought out from the studies by several workers.[53,54,67,81] Hutagalung found that when pigs were fed diets containing high levels of cassava root meal (60 to 75%) they developed diarrhea, skin lesions, localized swelling, and hind leg weakness, probably resulting from a zinc deficiency.[53] Iodine supplementation to cassava-based diets at the rate of 100 mg/kg was reported to have no effect on growth and feed efficiency.[118] Differences could be noticed in the carcass characteristics of these pigs.

A. Cassava Leaf Meal in the Nutrition of Swines

The potential of cassava leaf meal for feeding growing and finishing pigs was investigated by Alhassan and Odoi.[119] They found that 10% leaf meal was tolerated by the pigs without change in feed efficiency as compared with the control diet. Lee and Hutagalung found that the feeding of cassava leaves alone decreased the growth of pigs.[55] Supplementation of the diet with either sodium thiosulfate or methionine improved the performance. Molasses or palm oil addition to cassava leaf meal diets improved the performance still further.[120] No clinical symptoms of HCN poisoning were observed in pigs fed ad libitum with fresh cassava tops indicating its potential in swine feeding.[121] Rajaguru et al.[122] found that up to a 30% level, fattening pigs can tolerate cassava leaf meal. They have attributed this growth-promoting effect to the high lysine content of cassava leaf meal.

III. CASSAVA IN RUMINANT NUTRITION

Cassava is a potentially very useful feeding stuff for ruminants, mainly because of its bulky nature and high energy value. Among the feed components of cassava, the roots have the greatest nutritional value as they are the concentrated source of energy.

A. Milch Animals

Ahmed[123] found that the intake of dry grass could be reduced by supplementing the diet with fresh cassava roots. There was no variation in the pH of the rumen from the control and cassava fed animals. Fresh cassava roots have been reported to increase the milk yield up to 19.5% compared to control. Up to 10 kg of fresh cassava roots could be given per head per day without affecting the performance and quality of milk.[124] A large number of studies have been carried on the nutritive value of cassava as cattle feed. These studies in general show that cassava feeding increases milk yield.[125,126] Cassava tubers have been reported to yield untainted milk and butter of firm consistency.[127,128] It is reported to be efficiently utilized by dairy cows.[129,130]

Brouwer found that there was a slightly less milk yield and fat-free dietary methionine (DM) production in cassava-fed cows compared to corn control.[131] However, the fat percentage was 0.1% more in the milk from cassava-fed animals with the result that the total butter fat yield was similar for both groups. The effect of replacing oats with cassava at a 50 and 100% level was studied by Mathur et al.[132] and they found no adverse effect on the milk yield, fat percentage, etc. They suggested that the cereal portion of concentrate could be successfully replaced by cassava which made the ration economical. The replacement value of the mixture of cassava and groundnut on corn gluten was studied by Mallevre, who reported that the substitution was advantageous from an economical viewpoint.[133] The effect of feeding cassava, sweet potatoes, or edible canna as winter feed supplement for dairy cows was investigated by Assis.[134] He found that cassava increased milk yield by 19.5% compared to 7.8% for sweet potatoes. The response to cassava may be a result of the higher supply of metabolizable energy from cassava (11.9 to 14.6 MJ/kg) for cattle. Many of the studies on the replacement value of cassava indicate that it functions as an important source of energy in dairy cow rations.

B. Steers

Ahmed and Kay[135] found that the digestibility of crude fiber was significantly lower when either molasses or cassava was given with dried grass than when the grass was given alone. The effect of chopped cassava roots along with a protein supplement containing 12% urea and 75% cotton seed meal on the performance of crossbred steers was studied by Terleira et al.[136] They found that the beef grade was of the extra class for the cassava-fed steers and compared well with the animals fed control ration. On the other hand, Alquier has reported that cassava destroyed the food balance affecting the normal growth of cattle.[137] Castro and Desilva observed that although feed conversions improved with increased levels of cassava in steer rations, the daily weight gain and carcass weights decreased linearly with cassava inclusion percentage.[138] The effect of supplying molasses/urea with cassava chip meal on steer performance was studied by Garcia et al.[139] These studies revealed that replacing molasses by cassava meal was disadvantageous to the steers.

Variations in the protein quality of animal protein (fish meal) and vegetable protein (corn gluten meal) on the performance of steers fed cassava/urea meal were compared by Iriki et al.[140] Both the protein-supplemented groups performed better compared to cassava-urea control. However, plasma lysine and isoleucine were significantly higher for the fish meal supplemented cassava rations. The N-balance data indicated that N-retention was more for the vegetable protein supplemented group.

Although molasses is an almost essential ingredient in ruminant rations, its nonavailability in certain regions necessitated studies on the extent of replacing molasses with other ingredients such as whole sugar cane and cassava. Studies by Roverso et al.[141] showed that replacing molasses with cassava was superior to replacing with sugar cane. These results have later been confirmed by Rubio who found that daily weight gain of steers was 12.1% greater when 1 kg of molasses was replaced by 0.75 kg of cassava meal.[142] This indicated a favorable utilization of nitrogen by the urea in bacterial protein synthesis due to the presence of a complex carbohydrate (in cassava) that is degraded slower than sugar cane molasses. Teixeira[143] found that apparent digestibility of DM (excluding cellulose) and gross energy increased significantly when cassava supplemented with stylosanthes was fed to steers. Cellulose digestibility was more when cassava was supplemented with urea than with stylosanthes.

C. Growing Calves

Johnson et al.[144] found that calves fed on commercial and cassava-based diets grew faster than those fed on bran-based diets. The performance of young bulls was reported to be better by feeding 40% cassava meal compared to corn meal.[145] Cassava supplementation in the diets of calves has been reported to increase the daily weight gain, feed efficiency, and age at first service compared with Napier grass alone.[146] It was found that the best growth rates and efficiencies of feed conversion were obtained with 30% cassava inclusion. Studies by Mudgal and Sampath[147] showed that it was possible to replace either 50 or 100% of barley in the diets of growing calves with cassava. Later studies by Devendra and Lee[148] indicated the possibility of cassava inclusion up to 60% in the rations of growing calves. Although it is possible to substitute up to 50 to 60% of calf-starter diets, many studies reveal that this becomes economically feasible only when the price of cassava is less compared to corn.[149,150]

D. Goats and Sheep

Only very little studies have been reported on the feeding value of cassava in goats and sheep. Walker has reported the feeding of cassava peelings to goats and sheep.[151] Increased DM intake, apparent digestibility of dry matter, and N-balance were reported in castrated sheep fed napier grass silage with grated cassava added to it prior to ensiling.[152] Chicco et

Table 1
EFFECT OF FEEDING CASSAVA OR
MOLASSES ON UREA UTILIZATION
BY WEST AFRICAN WOOLLESS
SHEEP[153]

Parameters	Hay[a]	Hay + molasses	Hay + cassava
Body weight gain			
Male	88.3	106.7	100.9
Female	59.9	83.2	86.5
Bacterial protein (mg N/100 mℓ)	159.3	175.5	193.0
Blood urea (mg/100 mℓ)	15.6	19.8	20.7
Digestibility of organic matter (%)	59.6	70.7	66.6

[a] Hay basal diet contained 50% ground pangola grass,
   22% corn bran, 20% ground corn cobs, 5% molasses,
   2% urea, and 1% mineral mixture.

al.[153] found increased body weight gain and digestibility of organic matter in West African woolles sheep by feeding cassava along with hay. The results were almost comparable with the molasses fed group (Table 1).

Replacement of milled citrus pulp by cassava on the performance of castrated goats was studied by Akinsoyinu and Moa.[154] They found that any inclusion of cassava beyond 40% was not beneficial as the performance of the animals was only marginal. Castro and De Silva observed that the intake of dry matter, crude protein, and digestible protein by sheep decreased as the proportion of cassava was increased in the diet.[138] The apparent digestibility of dry matter and of gross energy increased linearly when maize was replaced by cassava. Devendra also found increased DM intake when cassava was supplemented in a maize diet.[155] Shultz et al.[156] obtained higher organic matter and cellulose digestibility in lambs fed cassava meal supplemented with sesame or uncooked urea or pressure cooked urea compared to the control fed a low quality hay. Extensive feeding trials conducted by Devendra indicated that maximum digestibility was obtained at 20% inclusion of cassava in sheep rations.[157]

There are only limited studies on the pattern of rumen fermentation consequent to cassava feeding. Chicco et al.[153] observed cassava feeding gave more propionic acid and less butyric acid than the molasses diet. This indicates that cassava feeding is energetically more advantageous to ruminants. The ruminal breakdown of starch is based principally on the proportion of amylolytic to nonamylolytic bacteria. High levels of cassava, therefore, lead to chances of incomplete ruminal breakdown of starch. Digestion of higher levels of cassava thus takes place considerably in the intestines posterior to the rumen. Rumen digestion of cassava can be facilitated by cooking cassava rather than feeding it to them raw uncooked.

E. Cassava Leaf Meal in Ruminant Nutrition

The nutritive value of cassava branches as a bran in dairy cattle feeding was investigated by Athanassof.[158] He concluded that cassava branches as bran was well accepted by cows at the rate of 10 kg/day/head. During drought seasons, when forage scarcity is experienced, this bran can find its place as a forage in ruminant diet. The in vivo digestibility of cassava canopy hay in sheep was studied by Barbosa.[159] The crude fiber digestibility was found to be reduced as the age of the leaves increased. The nutritive value of cassava leaves alone or in combination with pangola grass in sheep rations was studied by Bell and Norton.[160] They found that the addition of 33% cassava leaves significantly improved the voluntary feed intake of sheep. Devendra also found that cassava leaves were well tolerated by goats and sheep.[161] The apparent digestibility of DM, organic matter, and crude fiber was more for goats than sheep. Fernandez and Preston[162] observed that beef cattle could be economically fed on a molasses/urea diet in which the only source of fiber is cassava forage. The voluntary feed intake and total DM consumption could be significantly improved by feeding cassava forage along with sugar cane stems to bulls.[163] A higher final live weight was recorded in beef cattle fed elephant grass supplemented with cassava forage (25 to 50%). The use of cassava forage (0, 20, 40, to 60%) as a protein source in sugar cane/urea based diets for steers was assessed by Meyreles et al.[165] They concluded that the protein in cassava forage was easily soluble. There was also significant improvement in the voluntary feed intake in cassava forage fed steers. Teeluck et al.[166] fed weaned calves a basal diet comprising of molasses, urea, and fresh cassava forage and another set supplemented with cottonseed meal. They reported that supplementation of cotton seed meal along with cassava forage could improve the daily weight gain.

IV. NUTRITIONAL STUDIES ON SMALL ANIMALS

Several reports on nutritional evaluation of cassava on small animals pertain to the toxicity of cassava cyanogenic glucosides. However, few studies have been conducted on the performance of small animals fed cassava-rich diet. The growth curves of young offspring of guinea pigs fed cassava, sweet potatoes, and plantain were compared by Gale and Crawford.[167] They found uniform growth depression, poor fur and skin condition in all three experimental diets which they have attributed to the low fat content and amino acid imbalance of the foodstuffs. From their studies on rats, Ketiku and Oyenuga[168] found that the processing of cassava into gari was accompanied by a slight decrease in the gross energy from 4.5 to 4.4 kcal/g. The metabolizable energy values also indicated a decrease from 4.25 kcal/g for cassava to 3.69 kcal/g for gari.[168]

A 100% cassava diet fed to rats was reported to lead to injury of the osseous system and retarded bone growth and weight gain.[169] Supplementation of the diet with cod liver oil and casein was found to rectify the defects. Tasker observed that diet composed of 1:2 blend of corn and cassava did not promote any growth in albino rats.[170] The quality of the diet could be substantially improved by supplementing with a groundnut protein isolate and bengalgram flour, and this diet was found to significantly promote the growth of rats. They also recorded a higher fat content in the liver of rats fed a corn-cassava diet compared to those receiving protein foods. A moderate degree of parenchymal damage of the liver tissue resulting from protein deficiency from the cassava diet was also observed in these rats. The protein content of the carcass of the cassava-fed rats was much lower than those from the protein-food fed rats. This study thus confirms the importance of adequate protein supplementation on a high cassava diet.

Studies conducted by Fraser showed that cassava had no place in turkey diets.[171] The intestine of the turkey somehow found cassava incompatible and profuse watery diarrhea was a common feature in these birds. These animals later developed lameness and leg deformities which necessitated elimination of cassava from turkey diet by all compound feed manufacturers in Germany.

The efficacy of cassava flour as an ingredient of dried, pelleted fish feed was investigated by Jayaram and Shetty.[172] Better water stability of pelleted feed was observed with 14% incorporation of cassava. Ufodike and Matty[173] found cassava carbohydrate digestibility in fish was comparable to that of rice reaching up to 86 to 87%. They concluded that cassava could be substituted for rice as an energy source in carp diet.

Jeffers and Haep compared the nutritive value of some of the dehydrated tropical roots.[174] Satisfactory growth performance of rats was noted by feeding cassava, sweet potatoes, or yams up to a 20% level. Studies by Murthy et al.[175] showed that 25% of rice replaced by 4 parts of cassava and 1 part of groundnut cake flour could promote the growth of young rats. The biological value of cassava flour enriched with protein concentrates like fish, soybeans, and casein was worked out in mice by Nascimento.[176] It was found that rats fed on cassava flour supplemented with fish meal performed better compared to the other two groups. Orok and Bowland[177] observed that the nature of energy sources substantially influenced the response to protein supplementation and inclusion of cassava meal up to 50% was well tolerated by rats. Reports by Subrahmanyan et al.[178] also showed that substitution of rice, wheat, or ragi by cassava up to 25% gave distinct improvement in the growth response of rat. Further improvement could be produced by supplementing with groundnut cake flour. Substitution up to 50% by cassava flour did not deteriorate the overall growth-promoting value provided ad libitum diet was fed. The effect of fortification of cassava flour with soy flour on the performance of albino rats was studied by Temalilwa.[179] He found that 15% soy flour fortification was essential for the good performance of rats.

The nutritive value of cassava leaf protein concentrate (LPC) for rats was studied by Umoh.[180] It was observed that LPC was better utilized at levels up to 40% cassava. The supplementary role of LPC was further investigated by Umoh and Oke.[181] Although slightly inferior to casein, LPC can serve as a protein supplement to cassava rations. They found that with cornstarch, cassava LPC was a suitable source with high digestibility. The biological value and net protein utilization were 74 and 66% respectively in rats. Protein efficiency ratio (PER) was 1.8 and this decreased to 0.4 when 40% of maize was replaced with cassava. Umoh and Ayalogu[182] investigated the effect of different levels of palm oil and sulfur on the nutritive value of a cassava leaf protein diet in rats. The best results were obtained with 10% palm oil which could increase the biological value and net protein utilization and double the protein efficiency ratio (PER).

V. CASSAVA BY-PRODUCTS AS FEED SOURCE FOR ANIMALS

The rapid perishability of cassava tubers is a serious problem limiting the wide-scale utilization of cassava as human food and animal feed. This has been partially solved by converting fresh tubers into by-products such as chips, flour, starch, etc. which have comparatively prolonged shelf life. The utility of several by-products of cassava as an animal feed source has been studied by several workers.

The effect of modified cassava starch on the microflora of the small intestine and cecum of rats was studied by Bruns et al.[183] Hydroxypropyl distarch phosphate (HDP) and unmodified cassava starch were fed to two groups of rats. They found that the cecal lactobacilli increased in rats fed unmodified starch while constant counts were obtained in the HDP fed rats. The microbial data, however, indicated less digestibility of modified starch.

The effect of raw and cooked starches on hepatic enzyme activities was studied by Michaelis et al.[184] It was reported that the enzyme responses of rats given cooked cassava starch were similar to glucose-fed rats.

Fermentated cassava peel at levels of 20, 40, and 60% replacement of corn in diets composed of corn groundnut cake, brewer's dried grains, and cassava leaf meal was reported to give economic returns of 15, 15, and 19% respectively from sheep.[185] Cassava refuse obtained as a subproduct of the starch extraction units was assessed for its nutritive value as pig feed by Gomez et al.[186] It could be concluded from these studies that cassava refuse could be used as a cheap energy source for growing/finishing pigs, although it is slightly inferior to cassava meal. Penuliar[187] had reported earlier that cassava refuse was only 52% as efficient as rice bran as a swine feed and was suited as a feed source for older pigs.

The proximate analysis of cassava refuse (bran) is moisture 16%, crude protein 2.81%, fiber 6.57%, and starch 59.93%. Other carbohydrates and ash content are 12.01 and 1.89%, respectively. If properly balanced for protein, this can therefore find use as animal feed. Cassava chip flour waste could also be economically used in the rations at levels of 15 to 40% (dairy cows), 15 to 30% (work animals), 20 to 47% (growing pigs), 10 to 20% (breeding pigs), and 25 to 50% (fattening pigs). Lee and Yang[188] found that dried cassava pulp was efficiently utilized by pigs and the DM and crude protein digestibility were comparable to banana chips.

Cassava starch waste (thippi) could be economically substituted up to 20% without affecting the performance in the rations of piglets.[189] The nutritive value of cassava starch waste (CSW) supplemented with urea and molasses as a feed source for lambs was investigated by Reddy and Reddy.[190] Improved digestibility of DM, CP, and N-free extract was observed with CSW rations. Obioha and Anike studied the use of ensiled and sundried cassava peel on growing pig diets.[191] They observed that at 21.8% replacement of a maize diet, the inclusion of cassava peel reduced the feed cost of gain by 24%. Omole and Sonaiya[192] conducted studies on cassava peel meal utilization by rabbits and found that they could tolerate up to a 40% level when the diet was supplemented with fish meal. Supplementation with groundnut cake necessitated addition of 0.2% methionine and obtain equal performance with fish meal cassava peel diet. Sonaiya and Omole[193] also found that sundried cassava peel meal (CPM) promoted faster growth in pigs than in the control diet. The carcass quality was better for 15% CPM incorporated diet compared with lower levels.

VI. CASSAVA SILAGE AS ANIMAL FEED

The production of cassava silage has value as a means of preservation of cassava as feed material without altering the nutritional quality of the tubers even after long periods of storage. Early attempts in this line were made by Castillo et al.[194] who found that pigs fed cassava silage gained weights similar to those fed corn. Serres and Tillon[195] prepared cassava silage in big cylindrical cement vats and compared its quality with silages prepared in a ditch covered with a plastic sheet and in a metal vat. They found that the silage prepared in the cement vat was of superior quality and was readily acceptable by pigs.

The utilization of ensiled cassava in the rations of growing finishing pigs was investigated in detail by Buitrago et al.[196] The length of silage did not affect the daily weight gain of pigs. The feed efficiency of cassava silage was improved by the supplementation with salt, whereas the daily weight gain could be improved by supplementation with molasses. The effect of supplementation of cassava silage with protein sources like soybean cake, cotton seed cake, and fish meal was also studied by these workers. They found that vegetable protein sources were superior to fish meal when supplemented with cassava silage. Addition of salt at the time of ensiling reduced the consumption by the animals.[197] The use of cassava silage was found to reduce the feed costs in pig breeding.[198]

The use of cassava foliage silage as a forage for ruminants in periods of seasonal shortages of fresh fodder was studied by Carvalho et al.[199] They reported the inclusion of 25% cassava foliage with 75% elephant grass for maximum quality product. Cows weighing 400 kg have been reported to consume up to 22 kg of this silage per day. Up to 65% of the HCN was also eliminated in the ensiling process and hence, this should not pose any risk in the animals consuming the silage. Mixed cassava root elephant grass silage was attempted by Ferreira et al.[152] After 3 weeks of ensiling, the ensiled mass was fed to castrated sheep. The digestibility of dry matter and DM intake as well as N-balance could be significantly improved by the addition of grated cassava to elephant grass before ensiling. Ravelo et al.[200] observed that the sugar loss in sugar cane silage could be reduced by the addition of cassava forage and urea before ensiling.

Sprung[201] produced cassava root silage in laboratory fermentors and he found that there was no significant change in the protein content during the ensiling process. Silage made with added *Rhizopus oligosporus* did not result in increased protein levels. The protein content could be elevated to 6.5% by ensiling in thin layers and recycling the eluates to the solids. However, the process is not economical.

VII. MICROBIAL PROTEIN-ENRICHED CASSAVA FEEDS

The economic feasibility of cassava rations for animals depends mainly on the price of the cassava in relation to alternate energy sources and the price of supplementary protein sources to be added to satisfy the protein requirement of animals.[202] Because of the very low protein content of the cassava tubers, any substitution of cassava for cereals in compounded feeds necessitates the inclusion of huge amounts of supplementary proteins. Experimental studies conducted by Gomez et al. revealed that the life-cycle swine feeding program based on cassava meal requires approximately 60 to 65% more protein supplement than a similar feeding program using maize as an energy source.[95] Hence, in developing nations the potential of cassava and animal feed depends mainly on the availability of cheap protein sources. An alternate approach in this line is enriching cassava flour with microbial proteins. The microbial enrichment process is relatively cheap and the enriched product can elevate the status of cassava as a feed.

The concept of fermenting cassava for human consumption is not new as various fermented products of cassava such as peuyeum and gari are a part of the staple diet of Asians and Africans for many decades.[203-205] However, a large scale use of microorganisms for converting cassava carbohydrate to protein has started nearly 2 decades ago.

The use of fermented cassava in sheep rations by replacing corn up to 20, 40, and 60% was investigated by Adeyanju.[206] Digestibility of crude protein, crude fiber, N-free extract, etc. were significantly lower with 40 and 60% cassava. Dressing percentage, carcass length, and abdominal fat were also reduced with higher levels of cassava. Adeyanju and Pido studied the feeding value of fermented cassava peel in broiler diets.[207] Significant decreases in feed intake, daily weight gain, and feed efficiency ratio were noticed with increase in cassava levels. They observed that 20% was the optimum level at which fermented cassava peel could be included in broiler rations.

The nutritional evaluation of fermented cassava meal conducted by Ezeala[208] revealed that there is not much variation in the chemical constituents from the unfermented tuber meal. Fermented tuber meal had a 69% decrease in phosphorus content and 9.4% decrease in calcium content. They have suggested adequate supplementation of fermented tuber meal with minerals to obtain a response similar to the unfermented product. Fermented tuber meal protein had a 2% less biological vlaue compared with the unfermented meal.

Solid state fermentation of cassava with a *Rhizopus* spp. has been reported to elevate the protein level to 3.4%.[209] A microbial method for raising the protein level to 4% with minimum additives has been developed at Tropical Development and Research Institute (TDRI), London.[210] Solid state fermentation of cassava by microfungi to produce protein-enriched fermented cassava for animal feed has been attempted by Varghese et al.[211] They had also evaluated the role of natural nitrogenous supplements like chicken dung, pineapple bran, groundnut, etc. in advancing the fermentation of cassava. It was observed that direct fermentation of cassava with *Aspergillus, Neurospora,* and *Rhizopus* could elevate the protein values to 3%. Among the nitrogenous supplements tested, pineapple bran at 25% level elevated the protein to 4 to 5% while mixture of 12.5% pineapple bran and 12.5% chicken dung elevated the protein to 7%. Soybean and groundnut were found to be still better additives to facilitate protein enrichment.

Gray and Abou-El-Seoud[212] screened several filamentous fungi for SCP production and found that *Cladosporium eladosporoides* gave good mycelial yields.

The effect of substituting corn with fermented cassava in broiler rations was investigated by Hutagalung and Tan.[118] In their studies, they have used poultry manure as the natural nitrogen source in the ratio 1:3 (poultry manure:cassava). The true protein content of the fermented product was 10.0%. Ng and Hutagalung reported that supplementation of a 30% cassava with 15% poultry manure in broiler diets produced no adverse effects.[56] They also compared the nutritive value of fermented cassava with maize and found that the ash, crude fiber, crude protein, calcium, and phosphorus were higher in the former although fat and carbohydrate content were higher in the latter. Hutagalung and Tan[118] observed that the broiler fed fermented cassava compared well with those fed a corn diet. It was even seen that by replacing corn with 50% fermented cassava, the birds performed better compared with other treatments. Poor skin and egg yolk pigmentation were observed in the chickens fed fermented cassava compared to those on maize control.

A solid type fermentation process for cassava based on the traditional food processing techniques of Southern Asia was developed by TDRI, London.[213-215] The cassava flour enriched with ammonium nitrate, KH_2PO_4 and water is mixed with a fungus inoculum and the product is extruded into spaghetti-like strands. This is then allowed to ferment for 3 days. Out of the many fungal species tested only *Rhizopus* sp. was reported to elevate the protein content up to 2 to 4%. From the animal feed point of view, the product did not have much value as the process was too expensive and uneconomical.

Higher yields of protein have been reported in liquid fermentation systems based on cassava.[216] *Candida utilis* was allowed to grow on the cassava flour medium and the final product obtained was reported to contain 35% crude protein on a dry weight basis. Most yeasts are able to act on hydrolyzed starch only, and this necessitates the initial hydrolysis of starch by acids or enzymes. Filamentous fungi, on the other hand possess amylases and hence, can directly act upon starch. Brook et al.[213] obtained high protein yields with 27 strains of fungi and maximum yield of 33.6% crude protein was obtained from *Hetercephalum aurantiacum*. Gregory et al.[217] obtained very high crude protein yield (44%) by using *Aspergillus fumigatus* on cassava flour medium. A mutant of the strain *A. fumigatus* I-21A was extensively used for protein production from cassava by Gregory.[218] The conditions for the maximal yield of protein from cassava by this mutant were also standardized by him. The final yield of product from whole cassava was about 520 g/kg of carbohydrate supplied. The crude protein content of the dried product was 37%.

A major factor which needs attention while utilizing single cell protein (SCP) as animal feed component is the safety of the product. The finished product has to be entirely free from toxic hazards. The safety trials are usually conducted in small animals like rats, mice, etc. Khor et al.[219] found that the protein obtained with *A. fumigatus* I-21A mutant was harmless from rat feeding trials. The nutritive vlaue of microbial proteins obtained from cassava was assessed in rats by Alexander.[220] Analyses showed that microbial proteins were low in methionine and hence, supplementation with this amino acid was essential to give the protein a comparable status with animal proteins. Pilot plant trials of SCP production with the fungus *A. fumigatus* I-21A were undertaken by Santos and Gomez.[221] They obtained a slightly less crude protein content of about 28% than what was reported earlier from laboratory experiments.[217,222] The biological evaluation of the dried biomass showed that total weight gain of unsupplemented biomass was poor while methionine supplementation improved the PER values.

Farstad et al.[223] studied the effect of feeding "Pekilo" SCP to growing pigs given as a substitute to soybean protein. At 50% level, there was no negative effect on weight gain and carcass quality. But 100% substitution resulted in lower weight gain and feed efficiency.

The production of fungal biomass using cassava chips and leaves was studied by Sales

Table 2
AMYLASE ACTIVITY AND REDUCING SUGAR CHANGES DURING THE FERMENTATION OF CASSAVA STARCH

Culture	Amylase activity (hr)[a]				Reducing sugars (hr)[b]			
	24	48	72	96	24	48	72	96
R. chinensis	0.35	19.84	16.53	0.72	0.20	0.63	0.27	0.92
G. delequescence	9.38	23.34	8.04	0.90	0.27	0.89	0.30	0.90
C. eichhorniae	23.03	28.34	5.31	0.94	0.21	0.51	0.45	0.76
T. viride	0.84	34.48	29.76	1.03	0.16	0.84	0.18	0.51

[a] 1 unit = mg glucose liberated/hr/100 mg protein.
[b] Expressed as g/100 mℓ fermenting mash.

et al.[224] Out of seven fungal strains screened, *A. niger* and *Rhizopus* sp. produced biomass cocentrations above 18.5 g/ℓ. About 56 to 61% conversion of cassava carbohydrates was observed in submerged fermentation with these cultures and the protein content of dried biomass was 34%. In semisolid fermentation, the biomass protein obtained with *Rhizopus* sp. was only 9.64%. However, addition of dehydrated cassava leaves to the growth medium elevated the protein content to 13.74%.

The advantage of thermotolerant filamentous fungi for SCP production from cassava was investigated by many workers.[218,222,225,226] Toro[227] reported protein yields of 30% with *A. fumigatus* grown at 50°C in cassava pulp for 22 hr. The efficacy of two fungi *Rhizopus chinensis* 180 and *Cephalosporium eichhorniae* 152 in SCP production from cassava was investigated by Mikami et al.[228] Of these two fungal species tested *C. eichhorniae* was obligately acidophilic and hence, could not grow at the pH levels found in animal systems. As such, it poses relatively less health hazards for long-term trials with animals. The culture exhibited similar growth rates in fermentors containing cassava extract medium within the pH range 3.0 to 5.0. The optimum growth temperature was 45 to 47°C. The harvested biomass had 42.5% crude protein. The true protein content was 31%. The advantages of *C. eichhorniae* 152 over *A. fumigatus* I-21A are its superior performance in animal feeding trials and its greater safety.[229,230]

Fermentation of cassava starch for SCP production by four fungal cultures has been studied by Balagopalan and Padmaja.[231] The cultures tested were *R. chinensis, Gleocladium deleguescence, C. eichhorniae* and *Trichoderma viride*. Optimum temperature for growth was 45°C for *C. eichhorniae* and ambient temperature (30 ± 1°C) for the other three cultures. Samples were withdrawn from the fermentation broth at 24, 48, 72, and 96 hr after inoculation. Maximum amylase activity was observed in all four cultures at 48 hr fermentation (Table 2).

The mycelial mat was collected after 96 hr fermentation and maximum biomass yield was obtained with *C. eichhorniae* (Table 3). This indicates that 57% of cassava carbohydrate could be converted to protein by this culture. The crude protein yield was 50% with *C. eichhorniae* (Table 3). The other three cultures tested did not differ significantly in their ability for SCP production.

Gregory et al.[217] also reported similar yields of crude protein with *C. eichhorniae* (Table 4). However, *Rhizopus chinesis* has been reported to be a better converter of cassava carbohydrate to protein than reported from our studies.

They also found that cassava mash containing crude fiber was a poor substrate compared with cassava starch substrate. In the latter case, better utilization by *A. fumigatus* I-21A was noticed. In the context of the rapidly increasing price of conventional protein sources, the need for alternate nonconventional protein sources is becoming important. Single-cell pro-

Table 3
YIELD OF MYCELIA AND PROTEIN
DURING THE FERMENTATION OF
CASSAVA STARCH

Culture	Mycelial yield g/g carbohydrate supplied	Protein yield (%)
R. Chinensis	0.21	28.75
G. delequescence	0.14	26.25
C. eichhorniae	0.57	50.00
T. Viride	0.22	29.00

Table 4
NUTRITIONAL VALUE AND PROTEIN CONTENT
OF SELECTED THERMOTOLERANT FUNGI[217]

Fungi	Crude protein[a]	True protein[b]	PER[c]
Aspergillus fumigatus I-21	44.0	35.0	2.3
A. fumigatus I-21A	49.0	37.0	2.3
A. fumigatus I-21A (on cassava mash)	37.0	27.0	2.2
Cephalosporium eichhormae 152	49.0	38.0	2.6
Rhizopus chinersis 180	49.0	37.0	2.5

[a] Crude protein — N × 6.25.
[b] True protein — obtained after deleting amino N by assaying with bovine serum albumin.
[c] Protein efficiency ratio — g weight gain per g true protein consumed.

teins due to their relative case of preparation and economically beneficial nature will have a future in the animal feed industry utilizing cassava (a low protein crop) as an energy source.

REFERENCES

1. **Tabayoyon, T. T.,** The value of cassava refuse meal in the rations for growing chicks, *Philipp. Agric.,* 24, 509, 1935.
2. **Olson, D. W., Sunde, M. L., and Bird, H. R.,** The metabolizable content and feeding value of mandioca meal in diets for chicks, *Poult. Sci.,* 48, 1445, 1969.
3. **Maust, L. E., Scott, M. L., and Pond, W. G.,** The metabolizable energy of rice bran, cassava flour and black eye cowpeas for growing chicks, *Poult. Sci.,* 51, 1397, 1972.
4. **Muller, Z., Chou, K. C., and Nah, K. C.,** Cassava as a total substitute for cereals in livestock and poultry rations, *World Anim. Rev.,* 12, 19, 1974.
5. **Creswell, D. C.,** Cassava (*Manihot esculenta* Crantz) as a feed for pigs and poultry — a review, *Trop. Agric.,* 55, 273, 1978.
6. **Barrios, E. A. and Bressani, R.,** Composicion quimica de la raiz y de la hojas de algunas variedades de yuca Manihot, *Turrialba,* 17, 314, 1967.
7. **Vogt, H.,** The use of tapioca meal in poultry rations, *World's Poult. Sci. J.,* 22, 113, 1966.
8. **McMillan, A. M. and Dudley, F. J.,** Potato meal, tapioca meal and town waste in chicken rations, *Harper Adams Utility Poult. J.,* 26, 191, 1941.

9. **Klein, W. and Barlowen, V. G.**, Tapioca Mehl in Aufzuchfutter, *Arch. Geflugelk.*, 18, 415, 1954.

10. **Wegner, R. M.**, Zur Verbilligung von Kukenmastfuttermischungen, *Kraftfutter*, 44, 84, 1961.

11. **Vogt, H. and Penner, W.**, Der Einsatz Von Tapioca- und Maniokamehl in Geflügelmastfutter, *Arch. Geflügelk.*, 27, 431, 1963.

12. **Yoshida, M., Hoshu, H., Kosaka, K., and Morimoto, H.**, Nutritive value of various energy sources for poultry feed. IV. Estimations of available energy of cassava meal, *Jpn. Poult. Sci.*, 3, 29, 1966.

13. **Manjarrez, M. M., Fernandez, C. A. Cabrera, A. R., Carretero, M. A., Avita, E., and Shimada, A. S.**, Valor nutritivo de una combinacio'n de harina de yuca (*Manihot esculenta*) con puliduras de arroz como substituto del maizen la alimentacion de pollos y cerdos, in Investigaciones Sobre el Valor Alimenticiode la Yuca Para los Animales, Instituto National de Investigaciones Pecuwicas, Departmento de Divulgacion Tecnica, Mexico, 1973.

14. **Montilla, J. J., Mendez, C. R., and Wiedenhafer, H.**, Utilization de la harina de tubercuto de yuca (*Manihot esculenta*), en raciones inciadoras para pollos de engorde, *Arch. Latinoam. Nutr.*, 19, 381, 1969.

15. **Montilla, J. J., Mendez, C. R., and Wiedenhofer, H.**, Utilizacion de la harina de raiz de yuca *Manihot esculenta* en raciones para pollos de engorde, in *Commun. Cient. XIV Congr. Mundial Avicult.*, Madrid, Espana, 1970, 985.

16. **Vogt, H. and Stute, K.**, Prüfung von Tapiocapellets im Geflügelmast — Alleinfutter, *Arch. Geflügelk.*, 28, 342, 1964.

17. **Chou, K. C. and Muller, Z.**, Complete substitution of maize by tapioca in broiler rations, in *Proc. Aust. Poult. Sci. Convention*, World's Poultry Science Association, Auckland, New Zealand, 1972, 149.

18. **Phuah, C. H. and Hutagalung, R. I.**, Effect of levels of dietary protein and cassava on performance and body composition of chickens, *Malay. Agric. Res.*, 3, 99, 1974.

19. **Temperton, H. and Dudley, F. J.**, Tropical meal as a feed for laying hens, *Harper Adams Utility Poult. J.*, 26, 55, 1941.

20. **Falanghe, O.**, Substituicaodos farelos de trigo por farelos de arroz e mandioca na alimentacao de poedeiras, *Biologico*, 15, 35, 1949.

21. **Fangauf, R.**, Tapiokamhi im futter vom legehuhnern, *Futter Futtering*, 53, 426, 1955.

22. **Mantel, K.**, Effect of laying meals with different portions of cereals and the deficit of energy made up with cassava meal, *Arch. Geflügelk.*, 25, 373, 1961.

23. **Pillai, S. C., Srimath, E. G., Mathin, M. L., Naidu, P. M. N., and Muthana, P. G.**, Tapioca spent pulp as an ingredient in poultry feed, *Curr. Sci.*, 37, 603, 1968.

24. **Hamid, K. and Jalaludin, S.**, Utilization of tapioca in rations for laying poultry, *Malay. Agric. Res.*, 1, 48, 1972.

25. **Jalaludin, S. and Leong, S. K.**, Response of laying hens to low and high levels of tapioca, *Malay. Agric. Res.*, 2, 47, 1973.

26. **Eshiett, N. and Ademosun, A. A.**, Cassava for poultry, progress report on the use of cassava as animal feed in Nigeria, Ottawa, submitted to IDRC.

27. **Ademiosun, A. A. and Eshiett, N.**, Feeding cassava root meal to starter, grower and laying chickens, *Trop. Agric.* 57, 277, 1980.

28. **Enriquez, F. Q. and Ross, E.**, Cassava root meal in grower and layer diets, *Poult. Sci.*, 51, 228, 1972.

29. **Guimaraes, M. L. and Barros, M. S. C., de.**, *Bol. Tec. Div. Technol. Agr. Aliment. (Brazil)*, 4, 4, 1972.

30. **Agudu, E. W.**, Preliminary investigation on some unusual feedstuffs as yolk pigmenters in Ghana, *Ghana J. Agric. Sci.*, 5, 33, 1972.

31. **Ewing, R. A.**, *Poultry Nutrition*, 5th ed., Ray Ewing Publishing, Calif. 1963.

32. **Jensen, L. S., Fry, R. E., Alfred, J. B., and McGinnis, J. M.**, Improvement in the nutritional value of barley for chicks by enzyme supplementation, *Poult. Sci.*, 36, 919, 1957.

33. **Hildith, T. P. and Williams, P. N.**, *The Chemical Constitution of Natural Fats*, 4th ed., Chapman and Hall Ltd., London, 1964.

34. **Hudson, B. J. F. and Ogunsua, A. O.**, Lipids of cassava tubers (*Manihot esculenta* Crantz), *J. Sci. Food Agric.*, 25, 1503, 1974.

35. **Oyenuga, V. A.**, *Nigeria's Foods and Feeding Stuffs*, Ibadan University Press, Ibadan, 1968.

36. **Thornton, P. A. and Moreng, R. E.**, Further evidence on the value of ascorbic acid for maintenance and shell quality in warm environmental temperature, *Poult. Sci.*, 38, 594, 1959.

37. **Sullivan, T. W. and Kingan, J. R.**, Effect of dietary calcium level, calcium lactate and ascorbic acid on the egg production of White Leghorn hen, *Poult. Sci.*, 41, 1596, 1962.

38. **El-Boushy, A. R., Simons, P. C. M., and Wiertz, G.**, Structure and ultrastructure of the hen's egg shell as influenced by environmental temperature, humidity and vitamin C additions, *Poult. Sci.*, 47, 456, 1968.

39. **Hoyos, M. C. and Santos, N. J.**, Comparcion de dos niveles de harina de yuca con alto y bajo contenido de cianure en dietas para pollos de engorde, *Acta Agron.*, 33, 45, 1983.

40. **Huyghebaert, G. and de Groote, C.**, Vervanging van granen in legrantsoenen met varierendi voedings-densiteit verstrekt onder meel-en Korrelvora (Substitutos para cereales a differents concentraciones de nutrimentos, en mezcla o en comprimidos, para raciones de ponedoras), *Landbouwtijdschrift*, 36, 149, 1983.

41. **Khajarern, J., Khajarern, S., and Bunsiddhi, D.**, The substitutional equation in replacing cassava meal for maize in broiler rations. Cassava/nutrition project annual report, Khonkaen, Thailand, 1979, 57.

42. **Phalaraksh, K., Khajarern, J., and Puvadolphirod, S.**, An evaluation of the replacing value of cassava root meal for maize; broken rice or sorghum in layer diets, cassava/nutrition project, Annual Report, Khonkaen, Thailand, 1979, 85.

43. **Miranda, de R. M. Laun, G. F., and Costa, B. L.**, Emprego de marmelada de cavalo e de alfalfa em racoes de pintos, Rio de Janeiro Instituto de Zootecnia, 19, 1957, 18.

44. **Ross, E. and Enriquez, F. Q.**, The nutritive value of cassava leaf meal, *Poult. Sci.*, 48, 846, 1969.

45. **Siriwardene, J. A. de S. and Ranaweera, K. N. P.**, Manioc leaf meal in poultry diet, *Ceylon Vet. J.*, 22, 52, 1974.

46. **Rogers, D. J. and Milner, M.**, Amino acid profile of manioc leaf protein in relation to nutritive value, *Econ. Bot.*, 17, 211, 1963.

47. **Eggum, B. O.**, The protein quality of cassava leaves, *Br. J. Nutr.*, 24, 761, 1970.

48. **Enriquez, F. Q. and Ross, E.**, The value of cassava root meal for chicks, *Poult. Sci.*, 46, 622, 1967.

49. **Maner, J. H. and Gomez, G.**, Implications of cyanide toxicity in animal feeding studies using high cassava rations, in *Chronic Cassava Toxicity, Proc. Interdisciplinary Workshop*, Nestel, B. and Mac Intyre, R., Eds., London, IDRC, Ottawa, IDRC-010 e, 1973, 113.

50. **Maner, J. H.**, Management and feeding of pigs in the tropics, in *Animal production in the Tropics, Proc. Int. Symp Animal Production Trop.*, Heinemann Educational Books Nigeria, Ibadan, Nigeria, 1974.

51. **Job, T. A.**, Utilization and Protein Supplementation of Cassava for Animal Feeding and the Effects of Sulphur Sources of Cyanide Detoxification, Ph.D. thesis, Department of Animal Science, University of Ibadan, Nigeria, 1975.

52. **Babtunde, G. M., Fetuga, B. L., and Kassim, E.**, Methionine supplementation of low protein diets for broiler chicks in the tropics, *Br. Poult. Sci.*, 17, 463, 1976.

53. **Hutagalung, R. I.**, Nutritive Value of Leaf Meal, Tapioca Root Meal, Normal Maize and Opaque — 2 Maize and Pineapple Bran for Pig and Poultry, 17th Annu. Conf. Mal. Vet. Assoc., University of Malay, 1972, 1.

54. **Hew, V. F. and Hutagalung, R. I.**, The utilization of tapioca root meal (*Manihot utilissima*) in swine feeding, *Malay. Agric. Res.*, 1, 124, 1972.

55. **Lee, T. and Hutagalung, R. I.**, Nutritive value of tapioca leaf (*Manihot utilissima*) for swine, *J. Nutr.*, 40, 587, 1972.

56. **Ng. B. S. and Hutagalung, R. I.**, Evaluation of agricultural products and by-products as animal feeds. III. Influence of dehydrated poultry excreta supplementation in cassava diets on growth rate and feed utilisation of chickens, *Malay. Agric. Res.*, 3, 242, 1974.

57. **Hutagalung, R. I. and Chey, T. W.**, Evaluation of agricultural products and by-products as animal feeds. V. Utilization of pineapple bran and stearin by chickens, in *Proc. Conf. Malay. Food Self Sufficiency*, University of Malaya. Agric. Graduates Alumni, Malaya, 1975, 261.

58. **Hutagalung, R. J., Jalaludin, S., and Chang, C. C.**, Evaluation of agricultural products and by-products as animal feeds. II. Effects of levels of dietary cassava (tapioca) leaf and root on performance, digestibility and body composition of broiler chicks, *Malay. Agric. Res.*, 3, 49, 1974.

59. **Christensen, A. C., Knight, A. D., and Rauscher, G. F.**, An evaluation of cassava root meal as an energy source for broiler chicks, *Turrialba*, 27, 147, 1977.

60. **Tellez, B. G. and Caicedo, G. J.**, Addition of Vegetable Oil or Tallow to Broiler Diets Containing Cassava Meal from a Bitter Variety, thesis Zootecnista, Universidad National de Colombia, Palmira, 1983.

61. **Kroening, G. H., Pond, W. C., and Loosli, J. K.**, Dietary methionine — cystine requirement of the baby pig as affected by threonine and protein levels, *J. Anim. Sci.*, 24, 519, 1965.

62. **Berry, T. H., Combs, G. E., Wallace, H. E., and Robbins, R. C.**, Responses of the growing pigs to alterations in the amino acid pattern of isolated soyabean protein, *J. Anim. Sci.*, 25, 722, 1966.

63. **Luyken, R., de Groot, A. P., and Stratum van, P. G. C.**, Nutritional value of foods from New Guinea. II. Net protein utilization, digestibility and biological value of sweet potatoes, sweet potato leaves and cassava leaves from New Guinea, in utrecht, Central Institute for Nutrition and Food Research, 1961, 18.

64. **Muller, Z., Chou, K. C., and Nah, K. C.**, Cassava as a total substitute for cereals in livestock and poultry rations, in *Proc. Tropical Products Institute Conference*, 1975, 85.

65. **Tasker, P. K.**, Supplementary value of groundnut flour and blends of groundnut flour and skim milk powder to a maize tapioca diet, *Food Sci.*, 11, 181, 1962.

66. **Olson, D. W., Sunde, M. K., and Bird, H. R.**, The metabolizable content and feeding value of mandioca meal in diets for chicks, *Poult. Sci.*, 48, 1445, 1969.

67. **Hutagalung, R. I., Phuah, C. H., and Hew V. F.,** The utilization of cassava, tapioca (*Manihot utilissima*) in livestock feeding, in *Third Proc. Int. Symp. Trop. Root and Tuber Crops*, 1973, 45.

68. **Raymond, W. D., Jojo, N. Z., and Nicodemas, Z.,** The nutritive value of some Tanganyika foods. II. cassava, *East Afr. Agric. J.,* 6, 154, 1941.

69. **Oke, L.,** Chemical studies on some Nigerian foodstuffs "gari", *Nature (London),* 212, 1055, 1966.

70. **Savage, J. E., J. M., Pickett, E. E., and O'Dell, B. L.,** Zinc metabolism in the growing chick: tissue concentration and effect of phytate on absorption, *Poult. Sci.,* 43, 420, 1964.

71. **Edwards, H. M., Jr.,** The effect of protein source in the diet on Zn^{65} absorption and excretion by chickens, *Poult. Sci.,* 45, 412, 1966.

72. **Phuah, C. H. and Hutagalung, R. I.,** Zinc and iodine supplementation for chickens: effects of zinc and iodine supplementation in the cassava based diets on performance and body composition of broiler chickens, *Malay. Agric. J.,* 51, 311, 1978.

73. **Oberlease, D., Muhrer, M. E., and O'Dell, B. L.,** Dietary metal-complexing agents and zinc availability in rats, *J. Nutr.,* 90, 56, 1966.

74. **Arrington, L. R., Santa-Cruz, R. A., Harms, R. H., and Wilson, H. R.,** Effects of excess dietary iodine upon pullets and laying hens, *J. Nutr.,* 92, 325, 1967.

75. **Wilson, H. R., Fry, J. C., Harms, R. H., and Arrington, L. R.,** Performance of hens molted by various methods, *Poult. Sci.,* 46, 1406, 1967.

76. **Wilson, H. R. and Harms, R. H.,** High levels of dietary iodine and sexual maturity in males, *Poult. Sci.,* 51, 742, 1972.

77. **Harmon, B. G., Becker, D. E., Jansen, A. H., and Norton, H. W.,** Effects of Na_2 EDTA in diets of different iron levels on utilization of calcium, phosphorus and iron by rats, *J. Anim. Sci.,* 27, 418, 1968.

78. **Standish, J. F., Ammereman, C. B., Simpson, C. F., Neal, F. C., and Palmer, A. Z.,** Influence of graded levels of dietary iron, as ferrous sulphate on performance and tissue mineral composition of steers, *J. Anim. Sci.,* 101, 445, 1969.

79. **De Brochard, P., Rickard, J. R., and Triaca, H.,** Table international de composition des aliments, Marseilles, France, 1957.

80. **Chadha, Y. R.,** Sources of starch in Commonwealth territories. III. Cassava, *Trop. Sci.,* 3, 101, 1961.

81. **Muller, Z., Chou, K. C., Nah, K. C., and Tan, T. K.,** Study of nutritive value of tapioca in economic rations for growing finishing pigs, in *UNDP/SF Project SIN 67/505,* Vol. 672, Pig and Poultry Research and Training Institute, Singapore, 1972, chap. 1

82. **Carvalho, J. P. de., Monteiro, E. De. S., and Sources, L. M.,** Feijao macacar e raspa de mandipoca en substituicao ao farelo de trigo nas racoes de frangos de corte, in Brasil, Institute de Pesquisas Agronomicas de Pernambuco. Boletim Tecnico No. 39, 14, 1969.

83. **Hutagalung, R. I.,** Additives other than methionine in cassava diets, in *Cassava as Animal Feed, Proc. Workshop,* Nestel, B. and Graham, M., University of Guelph, Ontario, IDRC 095e, 1977, 18.

84. **Jalaludin, S. and Yin, O. S.,** Hydrocyanic acid tolerance of the hen, *Malay. Agric. Res.,* 1, 2, 1972.

85. **Yeong, S. W. and Syed, A.,** The use of tapioca in layer diets, *MARDI Res. Bull.,* 4, 91, 1976.

86. **Yeong, S. W. and Syed, A.,** Effects of sulphates and methionine supplementation in high cassava (*Manihot esculenta*) based diets for layers, *MARDI Res. Bull.,* 6, 202, 1978.

87. **Syed, A., Yeong, S. W., and Seet, C. P.,** Performance of layers fed high levels of broken rice and tapioca as a direct substitution for maize, *MARDI Res. Bull.,* 3, 63, 1975.

88. **Khajarern, S. and Khajarern, J. M.,** Use of cassava as a food supplement for broiler chicks, in *Proc. Fourth Symp. Int. Soc. Trop. Root Crops,* Cock, J., Mac Intyre, R., and Graham, M., Eds., IDRC, Ottawa, IDRC-080 e, 1976, 245.

89. **Zausch, M., Drauschke, M., and Lauterbach, A.,** Digestibility and use of cassava meal by pigs, *J. Tierernahr. Futterung,* 6, 256, 1968.

90. **Chicco, C. F., Garbati, S. T., Muller-Haye, B., and Vecchionacce, H. I.,** La harina de yuca en el engorde de cerdos, *Rev. Agron. Trop.,* 12, 599, 1972.

91. **Aumaitre, A.,** Valeur alimentaire du manioc et de differents cereales les regimes de sevrage pre' coce du porcelet: utilisation digestive de l' aliment et effect sur la croissance des animaux, *Ann. Zootech.,* 18(4), 385, 1969.

92. **Maust, L. E., Warner, R. G., Pond, W. G., and McDowell, R. E.,** Rice bran — cassava meal as a carbohydrate feed for growing pigs, *J. Anim. Sci.,* 29, 140, 1969.

93. **Maust, L. E., Pond, W. G., and Scott, M. L.,** Energy value of a cassava-rice bran diet with and without supplemental zinc for growing pigs, *J. Anim. Sci.,* 35, 953, 1972.

94. **Gomez, G., Camacho, C., and Maner, J. H.,** Utilization of cassava based diets in swine feeding, in *Proc. Fourth Symp. Int. Soc. Trop. Root Crops,* Cock, J., Mac Intyre, R., and Graham, M., Eds., IDRC, Ottawa, IDRC—080 e, 1977, 262.

95. **Gomez, G. G.,** Life-cycle swine feeding systems with cassava, *in Cassava as Animal Feed, Proc. Workshop,* Nestel, B. and Graham, M., Eds., University of Guelph, Ontario, IDRC-095 e, 1977, 65.

96. **Gomez., G. G., Santos, N. J., and Valdivieso, G. H.,** Use of cassava roots and products in animal nutrition, in *CIAT*, programa da Yuca. Yuca: investigacion, produccion y utilizacion, Cali, Colombia, 1982, 539.
97. **Castillo, L. S., Aglibut, F. B., Javier, T. A., Gerpacio, A. L., Garcia, G. V., Puyuoan, R. B., and Ramin, B. B.,** Camote and cassava tuber silage as replacement for corn in swine growing-fattening rations, *Philipp. Agric.*, 47, 460, 1963.
98. **Woodman, H. E., Menzies Kitchin, A. W., and Evans, R. F.,** The value of tapioca flour and sago pith meal in nutrition of swine, *J. Agric. Sci.*, 21(3), 526, 1932.
99. **Henry, Y.,** Effets nutritionnels de l'incorporation de cellulose purifice dans le regime du porcen croissance-finition, *Ann. Zootech.*, 19, 117, 1971.
100. **Hill, D. C.,** Chronic cyanide toxicity in domestic animals, in *Chronic Cassava Toxicity, Proc. Interdisciplinary Workshop*, Nestel, B. and Mac Intyre, R., Eds., London, IDRC-010 e, 1973, 105.
101. **Acuna, C. A.,** Eficuncia de la harina de yuca (*Manihot esculenta*) descortexada como substituto parcial del maize en el engorde de cerdos, *Oriente Agropecu.*, 3 (1/2), 24, 1971.
102. **Alvarez, G. R. and Alvarado, R. L.,** La yuca (*Manihot esculenta* Crantz) como fuente energetica en la alimentacion de los cerdos. II. Substitucion del maiz por custro niveles de harina de yuca deshidratada en raciones para cerdos en crecimiento, *Ganagringo*, 10(40), 42, 1975.
103. **Arambawela, W. J. et al.,** Effect of replacing barley with tapioca meal at two different levels of feeding on the growth and health of early weaned pigs, *Livestock Prod. Sci.*, 2, 281, 1975.
104. **Asico, P. M.,** A comparative study of gaplek meal and corn as basal feed for growing and fattening pigs, *Philipp. Agric.*, 29, 706, 1941.
105. **Barbosa, A. S. et al.,** Dried cassava as a substitute for wheat by-products in fattening swine, *Arguivos Escola Superior Vet. (Brazil)*, 10, 15, 1957.
106. **Boon, O. C.,** A comparison of Sago flour versus tapioca root meal with varying proportions of fish meal and soyabean meal for growing/finishing pigs, *Malay. Agric. J.*, 50, 427, 1976.
107. **Fullerton, J.,** Tapioca meal as food for pigs, *J. Minist. Agric. (England)*, 36, 130, 1929.
108. **Gomez, G. and Santos, J.,** Evaluacion nutritiva de la harina de yuca con y sin melaza como fuente energetica para cerdos en crecimiento y acabado. In Ruunion Latinoamericana de Produccion Animal, 7a, Panama, 1979, Programa y compendios, Panama, Association, Panamena de produccion Animal PP NR 48. Esp
109. **Hansen, V., Bresson, S., and Jensen, A.,** Tapioka melsom foder til slagterisvin. (La harina integral de yuca como alimento para cerdos tipo magro.) Kobenhavn Denmark. Beretning fra statens Husdyrbrugsfors g. No. 440, 20, 1976.
110. **Hofman, P.,** Cassava meal for fattening swine, Bavaria Landesanstalt für Tierzucht in Brub, *Milteilungen.* 8, 32, 1960.
111. **Kitpanit, N. and Bunsiddhi, D.,** The substitutional value of cassava root product of various quality grades for rice by-products in semi-commercial pig rations, in *Cassava/Nutrition Project Annual Report 1979*, Khon Kaen, Thailand, 1979, 95.
112. **Peixoto, R. R. and Farias, J. V. Da S.,** Estudo da influencia da prensagem (pellets) de racao com elevado teor de farinha de mandioca no desempento de porcos em crescimento e terminacao, e nas caracteristicas da carcaca, *Rev. Soc. Bras. Zootecnia*, 5, I, 1976.
113. **Totsuka, K. et al.,** Study of nutritive value of cassava in rations for growing and finishing pigs, *Jpn. J. Zootech. Sci.*, 49, 250, 1978.
114. **Zoby, J. L. E., Campos, J., Mayrose, V., Melgact, P., and Costa, A.,** Cassava meal fat and methionine supplemented in swine feeding, *Rev. Ceres*, 18, 195, 1971.
115. **Devendra, C. and Hew, V. F.,** The utilisation of varying levels of dietary palm oil by growing finishing pigs, *MARDI Res. Bull.*, 5, 83, 1977.
116. **Shimada, A. S., Peraza, C., and Cabello, F. T.,** Valor alimenticio de la harina de yuca *Manihot utilissima Pohl, Para cerops, Tec. Pecu. Mex.*, 15, 31, 1971.
117. **Hew, V. F. and Hutagalung, R. I.,** Utilization of cassava as a carbohydrate source for pigs, in *Proc. Fourth Symp. Soc. Trop. Root Crops*, Cock, J., MacIntyre, R., and Graham, M., Eds., Ottawa, IDRC-080 e, 14, 1976, 242.
118. **Hutagalung, R. I. and Tan, P. H.,** Utilization of nutritionally improved cassava in poultry and pigs diet, in *Proc. Fourth Symp. Int. Soc. Trop. Root Crops*, Cock, J., MacIntyre, R., and Graham, M., Eds., IDRC, Ottawa, IDRC—080 e, 1976, 255.
119. **Alhassan, W. S. and Odoi, F.,** Use of cassava leaf meal in diets for pigs in humid tropics, *Trop. Anim. Health Prod.*, 14, 216, 1982.
120. **Lee, T.,** The Nutritive Value and Utilization of Tapioca Leaf on the Performance of Swine, thesis B. Agric. Sci., Faculty of Agriculture, University of Malaya, Kuala Lumpur, 1972.
121. **Mahendranatha, T.,** The effects of feeding tapioca (*Manihot utilissima* pohl) leaves to pigs, *Malay. Agric, J.*, 48, 60, 1971.

122. **Rajaguru, A. S. B., Ravindran, V., and Ranaweera Banda, R. M.,** Manioc leaf meal (*Manihot esculenta* Crantz) as a source of protein for fattening swine, *J. Natl. Sci. Counc. Sri Lanka*, 7, 105, 1979.

123. **Ahmed, F. A.,** Feeding cassava to cattle as an energy supplement to dried grass, *East Afr. Agric. For. J.*, 42, 368, 1977.

124. **Athanassof, N.,** Mandioca para as vaccas leiteiras, *Chacaras Quintaes*, 12, 455, 1915.

125. **Cardoso, R. M., Campos, J., Hill, D. H., and Coelho, J. F., de Silva,** Effeito de substituticao gradativa do milho pela raspa de mandioca, na producao de leite, *Rev. Ceres*, 14, 308, 1968.

126. **Olaluku, E. A., Egbuiwe, A. M., and Oyenuga, V. A.,** The influence of cassava in the production ration on the yield and composition of white Fulani cattle, *Niger. Agric. J.*, 8, 36, 1971.

127. **Pyanaert, L.,** *Le Manioc*, 2nd ed., Ministere des colonies, Bruxelles, 1951.

128. **Brouwer, E.,** Feeding test with tapioca meat on milk cows, 1931, Annual Report, 1933, 29.

129. **Mandioca, E.,** Batata-doce aumentam leit na reca, *Coopercotia*, 19, 52, 1969.

130. **Albuque, de M.,** Estado atual das pesquisas com mandioca no IPEAN in Runiao de Comissano Nacional de Mandioca. 5a, Sele Lagos, Minas Gerain, Brazil, 1971.

131. **Brouwer, E.,** Cassava meal for dairy cattle, in *Vereeniging tot exploitatie eener proefzui velboerdery te Hoorn*, Hoorn, Holland, A. Houdik, 1932, 79.

132. **Mathur, M. L., Sampath, S. R., and Ghosh, S. N.,** Studies on tapioca: effect of 50 and 100 per cent replacement of oats by tapioca in the concentrate mixture of dairy cows, *Indian J. Dairy Sci.*, 22, 193, 1969.

133. **Mallevre, M.,** Cassava meal in feeding dairy cows, Bulletin des Seances de la Societe Nationale d' Agriculture de France, 74, 638, 1914.

134. **Assis, F de P.,** Efeitos da administracao de raizes e tuberculos, como suplements de inverno, na alimentacao de vacas em lactacao, *Bol. Indust. Anim.*, 20, 55, 1962.

135. **Ahmed, F. A. and Kay, M. A.,** A note on the value of molasses and tapioca as energy supplements to forage for growing steers, *Anim. Prod.*, 21, 191, 1975.

136. **Terleira, G. H., Tenbrinke, H. W., Lopez., C. W., and Santisteban, S. D.,** Uso de raices de yuca, coronta de maiz y cascara de algodon en el engorde de novilos en Tarapota — San Martil Lima, Peru, Ministerio de Alimentacion, Direccion General de Investigacion, 1975, 13.

137. **Alquier, J.,** Valeurs nutritives comparees, pourles bovins, des gros sons de ble, des issues roses de riz et de la mouture de manioc. Bulletin de la Societe Scientifique d'Hygiene Alimentaire et d'Aoimentation Rationnelle de l' Homme, 15, 294, 1927.

138. **Castro, M. E. D. and Silva, J. F. C. Dl.,** La Larins integral de yuca como substituto de la mazorca de maiz molida. II. Bovinos en confinamiento *Experiential*, 20, 204, 1975.

139. **Garcia, J. A., Campos, J., and Peres, F. L.,** Melazal urea × harina integral de yuca/urea en el engorde de bovinos en confinamiento, *Sieva* 30, 9, 1970.

140. **Irik, T., Shibui, H., and Abe, M.,** Effects of supplemental fish meal and corn gluten meal in a diet containing urea and cassava on the growth and the nitrogen retention of steers, *Jpn. J. Zootech. Sci.*, 48, 748, 1977.

141. **Roverso, E. A., Tundisi, A. G. A., and Lima, F. P.,** Molasses, cassava and whole sugar cane in rations for Nelore bovines, *Rev. Med. Vet.*, 5, 36, 1969.

142. **Rubio, C. E.,** Effetco comparativo de la melaza de cana y harina de yuca en la utilization de urea en la alimentacion de rumiantes, *Rev. ICA.*, 13, 537, 1978.

143. **Teixeira, L. B.,** Urea, Stylosanthes Y harina integral de yuca como suplementos de Pennisetum purpureum para bavinos en confinamiento. Tese Mag. Sc. Vicosa M. G., Universidade Federal de Vicosa, 1975.

144. **Johnson, P. T. C., Rose, C. J., and Mills, W. R.,** Nutritional studies with early weaned beef calves, *Rhod. J. Agric. Res.*, 6(1), 5, 1968.

145. **Montilla, J.,** La raiz de en la alimentation animal, *Rev. Protinal*, 18, 120, 1971.

146. **Devendra, C. and Lee, T.,** Studies on Kedah, Kelantan cattle. I. The effect of improved nutrition on growth, *MARDI Res. Bull.*, 3, 68, 1975.

147. **Mudgal, V. D. and Sampath, S. R.** Review of research work carried out in animal nutrition, in *National Dairy Research Institute*, Karnal, India, 1972, 68.

148. **Devendra, C. and Lee, T.,** Studies on Kedah-Kelantan cattle. II. The effect of feeding increasing levels of tapioca (*Manihot utilissima* Pohl), *MARDI Res. Bull.*, 4, 80, 1976.

149. **Peixoto, R. R.,** Utilizacion de harina integral de yuca en raciones y consideraciones economicas sobre el destete precoz de terneros holandeses. Pelotas, Rio Grande do Sul, Brazil Universidade Federal de Pelotas, *Indicacao Pesquisa*, 2, 3, 1974.

150. **Valdivieso, C. A.,** Comparative Study of Cassava and Corn Meals as Feed Mixtures for Weanling Calves, thesis Mag. Agr. *Turrialba*, Costa Rica IICA, 1958.

151. **Walker, A.,** Un aliment de famine: l'ecorce de manioca, *Rev. Int. Bot. Appl.*, 31, 542, 1951.

152. **Ferreira, J. J., Silva, J. F., Da., and Gomide, J. A.,** Efeito do estadio de desenvolvimento, do emurcheumento e da adicao de raspa de mandioca sobre o valor nutritivo da silegem de capim-elephante (*Pennisetum purpureum* Schum), *Experiential*, 17(5), 85, 1974.

153. **Chicco, C. F., Carnevalia, A., Shultz, T. A., Shultz, E., and Ammerman, C. B.,** Cassava and molasses in the utilization of urea by feeding lambs, *Mem. Assoc. Latin. Prod. Anim.,* 6, 7, 1971.

154. **Akinsoyinu, A. O. and Moa, A. U.,** Influence of dietary levels of cassava flour on dried citrus pulp utilization by the West African dwarf goats, *Afr. J. Agric. Sci.,* 5(2), 41, 1978.

155. **Devendra, C.,** Studies on the utilization of rice straw by sheep. IV. Effect of carbohydrate source on the utilization of dietary urea and nitrogen retention, *Malay. Agri. J.,* 50, 358, 1976.

156. **Shultz, T. A. Shultz, E., and Chicco, C. F.,** Pressure cooked urea-cassava meal for lambs consuming low quality hay, *J. Anim. Sci.,* 35, 865, 1972.

157. **Devendra, C.,** Cassava as a feed source for ruminants, in *Cassava as Animal Feed Proc. Workshop,* Nestel, B. and Mac Intyre, R., Eds., IDRC, Ottawa, IDRC-095 e, 1977, 107.

158. **Athanassof, N.,** Contribucion para el estudio de ramas de yuca como forraje en la alimentacion da Agricultura, Commercio e obras Publicas do Estado de Sao Paulo, 25, 1923.

159. **Barbosa, C.,** Aprovechaimento de la parte aerea de yuca en la alimentacion animal. Tese Mag. Sci. Piracicaba-SP, Brasil, Escola Superior de Agricultura Luiz de Queiroz da Universidade de Sao Paulo, Brasil, 71, 1972.

160. **Bell, G. D. and Norton, B. W.,** The nutritive value of cassava (*Manihot esculenta*) leaf for sheep, in *Cassava Research Program, University of Queensland,* St. Lucia, Australia, 1981, 84.

161. **Devendra, C.,** The utilization of forages from cassava, pigeon pea, leucaena and groundnut by goats and sheep in Malaysia, in Nutrition Systemes d'Alimentation International Vol. 1, Morand-Fehr, P., Bourobouze, A., Simiane, M. de, Eds, Tours, France, 1981, 338.

162. **Fernandez, A. and Preston, T. R.,** Forraje de yuca Como suplemento de fibra y proteina en dietas de melaza: efecto del nivel de forraje y supplementacion con harina de soya, *Prod. Anim. Trop.,* 3, 111, 1978.

163. **Efoukkes, D. and Preston, T. R.,** Efecto sobre el consumo voluntario y digestibilidad de supplementar tallo de canan integral picado con puntas de cana hyas de platano o forraje de yuca, *Prod. Anim. Trop.,* 4, 36, 1979.

164. **Guzman, M. R., De.,** Integration of backyard dairy beef farming with the cropping systems of South East Asia. Taipei, Taiwan, Asian and Pacific council, Food and Fertilizer Technology centre. Extension Bulletin No. 110, 18, 1978.

165. **Meyreles, I., Macleod, N. A., and Preston, T. R.,** Forraje de yuca como fuente proteica en dietas de cana de azucar para el ganado: efecto de diferentes niveles de yuca y urea sobre parametros de fermentacion ruminal, *Prod. Anim. Trop.,* 2, 309, 1978.

166. **Teeluck, J. P., Nicolin, R., Hulman, B., and Preston, T. R.,** Apuntes sobre el uso de la yuca (*Manihot esculenta*) como fuente combinada de proteina y forraje para el crecimiento de becerros alimentados con dietas de melaza/urea, *Prod. Anim. Trop.,* 6, 90, 1981.

167. **Gale, M. M. and Crawford, M. A.,** The effect of African staple foodstuffs on guinea-pig growth curves, *Trans. Soc. Trop. Med. Hyg.,* 63, 821, 1969.

168. **Ketiku, A. O. and Oyenuga, V. A.,** Dehydrated yam (*Dioscorea rotundata,* Poir) and cassava (*Manihot utilissima,* Pohl) as sources of energy to the laboratory rat, *West Afr. J. Biol. Appl. Chem.,* 16, 9, 1973.

169. **Martino, G. and Chenu-Bordon, J. G.,** El valor alimienticio de la yuca, *Arch. Sci. Biol.,* 17, 305, 1932.

170. **Tasker, P. K.,** Supplementary value of groundnut flour and blends of groundnut flour and skim milk powder to a maize-tapioca diet, *Food Sci.,* 11, 181, 1962.

171. **Fraser, D. M. K.,** Manioc unsuitable for turkeys, *Vet. Rec.,* 93, 238, 1973.

172. **Jayaram, M. G. and Shetty, H. P. C.,** Formulation, processing and water stability of two new pelleted fish feeds, *Aquaculture,* 23, 355, 1981.

173. **Ufodike, E. B. C. and Matty, A. J.,** Growth responses and nutrient digestibility in mirror carp (*Cyprinus Carpio*) fed different levels of cassava and rice, *Aquaculture,* 31(1), 41, 1983.

174. **Jeffers, H. F. K. and Haep, H.,** A preliminary study of the nutritive value of some dehydrated tropical roots, in *Proc. Inst. Int. Symp. Trop. Root Crops,* University of West Indies, Augustine, Trinidad, 2, 72, 1969.

175. **Murthy, H. B. N., Swaminathan, M., and Subrahmanyan, V.,** Supplementary value of groundnut cake to tapioca and sweet potato, *J. Sci. Indian Res.* 9B, 173, 1950.

176. **Nascimento, J. S.,** Descripcion de los trabajos experimentales realizados en el Departmento de Nutricion Experimental con harina de yuca, in *Reuniao da Comissao Nacional da Mandioca,* Vol. 6a, Recife, Pernambuco, Brasil, 1972, 31.

177. **Orok, E. J. and Bowland, J. P.,** Nigerian Cocoa Husks and cassava meal as sources of energy for rats fed soybean meal or peanut meal supplemented diets, *Can. J. Anim. Sci.,* 54(2), 229, 1974.

178. **Subrahmanyan, V. et al.,** Investigations on grains substitutes. V. The nutritive value of synthetic rice. Bulletin, Central Food Technological Research Institute (India), 4(3), 55, 1954.

179. **Temalilwa, C. R.,** Cassava (*Manihot esculenta* Crantz) Flour Fortification with Soybean Flour, Mag. Sci. thesis, University of Tennessee, Knoxville, 1980.

180. **Umoh, I. B.,** Effect of levels of cassava on the utilization of leaf protein in the rat, *Nutr. Rep. Intl.* 16, 397, 1977.

181. **Umoh, I. B. and Oke. O.,** The supplementary role of leaf protein concentrate in some tropical and subtropical food in rats, *Nutr. Rep. Inter.,* 16, 29, 1977.

182. **Umoh, I. B. and Ayologu, E. O.,** Effects of different levels of palm oil and sulphur in cassava-based diets, *Food Chem.,* 10, 83, 1983.

183. **Bruns, P., Hood, L. F., and Seeley, H. W., Jr.,** Effect of modified starch on the microflora of the small intestine and caecum of rats, *Nutr. Rep. Inter.,* 15, 131, 1977.

184. **Michaelis, O. E., Nace, C. S., and Szepesi, B.,** Effect of refeeding raw and cooked starches on hepatic enzyme activities of rats, *Br. J. Nutr.,* 39, 85, 1978.

185. **Adebowale, E. A.,** Reemplazamiento del maiz por cascaras de la yuca fermentada (*Manihot* utilissima Pohl) on raciones para ovejas, *Prod. Anim. Trop.,* 6, 58, 1981.

186. **Gomez, G., Santoz, J., and Taborda, E.,** utilizacion de ripio o afrecho de yuca en dietas para cerdos en crecimiento y acabado, in *Reunion Latinoamericana de Production Animal,* 7 a, Panama, 1979.

187. **Penuliar, S. P.,** A comparative study of cassava refuse meal and rice bran as feeds for growing and fattening pigs, *Philipp. Agric.,* 29, 611, 1940.

188. **Lee, P. K. and Yang, Y. F.,** Estudio comparative de la digestibilidad nutritiva de trozos de batata, bagazo seco de yuca y trozos de banano en porcinos, *J. Agr. Assoc. China,* 102, 96, 1978.

189. **Manickam, R. and Gopalakrishnan, C. A.,** Studies on feeding starch waste to swine, *Indian J. Anim. Res.,* 12, 13, 1978.

190. **Reddy, T. K. and Reddy, M. R.,** Studies on the utilization of urea-molasses enriched paddy straw and tapioca residue in lamb rations, *Indian Vet. J.,* 56, 400, 1979.

191. **Obioha, F. C. and Anikwe, P. C. N.,** Utilization of ensiled and sundried cassava peels for growing swine, *Nutr. Rep. Int.,* 26, 961, 1982.

192. **Omole, T. A. and Sonaiya, E. B.** The effect of protein source and methionine supplementation on cassava peel meal utilization by growing rabbits, *Nutr. Rep. Int.,* 23, 729, 1981.

193. **Sonaiya, E. B. and Omole, T. A.,** Cassava peels for finishing pigs, *Nutr. Rep. Int.,* 16, 479, 1977.

194. **Castillo, L. S., Aglibut, F. B., Javier, T. A., Gerpacio, A. L., Puyuoan, R. B., and Ramin, B. B.,** Camote and cassava tuber silage as replacement for corn in swine growing fattening rations, *Philipp. Agric.,* 47, 460, 1964.

195. **Serres, H. and Tillon, J. P.,** L'ensilage des racines de manioc, *Rev. Elev. Med. Vet. Pays Trop.,* 25(3), 455, 1972.

196. **Buitrago, J., Santos, J., and Gomez., G.,** Ensilado de raices de yuca en raciones para cerdos en crecimiento y acabado, in *Reunion Latinoamericana de Production Animal,* 7a, Panama, 54, 1979.

197. **Conci, V. A. and Leboute, E. M.,** supplemetacion con ensilaje de raices de yuca en engorde y acabado de cerdos *Anuario Tecnico do Instituto de Pesquisas, Zootec. Francisco Osorio,* 5(2) 1023, 1978.

198. **Garcia, I. M. D. and Leboute, E. M.,** Raices de yuca como fuente energetica en raciones de cerdos en recria y acabado. *Anuario Tecnico do Institute de Pesquisas, Zootec. Francisco Osorio,* 5(2), 1023, 1078.

199. **Carvalhe, J. L. H. De., Perim, S., Costa, I. R. S.,** Parte Aerea de la yuca en alimentacion animal. I. Valor nutritive y calidad de ensilage. Centro de Pesquisa Agropecuar ia dos cerrados, *Commun. Tech.,* 29, 6, 1983.

200. **Revelo, G., Macleod, N. A., and Preston, T. R.,** Ensilaje de la cana de azucar, Forraje de yuca y urea, *Prod. Anim. Trop.,* 2, 34, 1977.

201. **Sprung, D. W.,** Improvement of the Nutritional Value of Cassava by the use of High-Solids Fermentations, Master Science thesis, University of Guelph, Canada, 1974.

202. **Phillips, T. P.,** *Cassava utilization and Potential Markets,* International Development Research Centre, Ottawa, IDRC-020 e 2, 1974.

203. **Herseltine, C. W.,** A millenium of fungi, food and fermentation, *Mycologia,* 57, 149, 1965.

204. **Gray, W. D.,** The use of fungi as food and in food processing, *CRC Crit. Rev. Food. Technol.,* 1, 225, 1970.

205. **Pederson, C. S.,** *Microbiology of food fermentation,* AVI Publishing, Westport, Conn., 1971.

206. **Adeyanju, S. A.,** Maize replacement value of fermented cassava in rations for sheep, *Turrialba,* 29, 1979.

207. **Adeyanju, S. A. and Pido, P. P.,** The feeding value of fermented cassava peel in broiler diets, *Nutr. Rep. Int.,* 18, 79, 1978.

208. **Ezeala, D. O.,** Changes in the nutritional quality of fermented cassava tuber meal, *J. Agric. Food Chem.,* 32, 467, 1984.

209. **Yeang, H. Y.,** Protein enrichment of tapioca by fermentation, Preliminary fermentation work report (Communication to IDRC) Univ. Malaya, Kuala Lumpur, Malaysia, 1973.

210. **Woollen, A. H.,** What's new in Europe among advance *Food Eng.,* 40, 98, 1968.

211. **Varghese, G., Thambirajah, J. J., and Wong, F. M.,** Protein enrichment of cassava by fermentation with microfungi and the role of natural nitrogenous supplements, in *Proc. Fourth Symp. Int. Soc. Trop. Root Crops,* Cock., J., MacIntyre, R., and Graham, M., Eds., IDRC, Ottawa, IDRC-080 e, 1976, 250.

212. **Gray, W. D. and Abou-et-Seoud, M. O.,** Fungal Protein for food, and feeds. III. Manioc as a potential crude raw material for tropical areas, *Econ. Bot.,* 20, 251, 1966.

213. **Brook, E. J., Stanton, W. R., and Wallbridge, A.,** Fermentation methods for protein enrichment of cassava, *Biotechnol. Bioeng.,* 11, 1271, 1969.

214. **Stanton, W. R. and Wallbridge, A.,** Fermented food processes, *Proc. Biochem.,* 4, 45, 1969.

215. **Stanton, W. R. and Wallbridge, A.,** Improvements relating to the fermentation of cassava and other vegetables substances, British Patent, 1, 277, 002, 1972.

216. **Strasser, J., Abbett, J. A., and Battey, R. F.,** Process enriches cassava with protein, *Food Eng.,* 42, 112, 1970.

217. **Gregory, K. F., Reade, A. E., Khor, G. L., Alexander, J. C., Lumsden, J. H., and Losos, G.,** Conversion of carbohydrates to protein by high temperature fungi, *Food Technol.,* 30(3), 30, 1976.

218. **Gregory, K. F.,** Cassava as a substrate for single cell protein production: microbiological aspects, in *Cassava as Animal Feed, Proc. Workshop,* Nestel, B. and Graham, M., Eds., IDRC, Ottawa, IDRC-095 e, 1977, 72.

219. **Khor, G. L., Alexander, J. C., Reade, A. E., Gregory, K. F., Lumsden, J. H., and Losos, G. J.,** Nutritional and safety evaluations of microbial proteins from cassava, *Proc. Tenth Int. Nutrition,* Kyoto, Japan, 1975.

220. **Alexander, J. C.,** Nutrition of laboratory animals with fungi grown on cassava, in *Cassava as Animal Feed, Proc. Workshop,* Nestel, B. and Graham, M., IDRC, Ottawa, IDRC-095 e, 1977, 85.

221. **Santos, N. J. and Gomez., G.,** Pilot plant for single cell protein production, in *Cassava as Animal Feed Proc. Workshop,* Nestel, B. and Graham, M., Eds., IDRC, Ottawa, IDRC-095 e, 1977, 91.

222. **Reade, A. E. and Gregory, K. F.,** High temperature production of protein-enriched feed from cassava by fungi, *Appl. Microbiol.,* 30(16), 897, 1975.

223. **Farstad, L. et al.,** Effects of feeding ''Pekilo'' single cell protein in various concentrations to growing pigs, *Acta Agric. Scand.,* 25, 291, 1975.

224. **Sales, A. M., Menezes, T. J. B. De. Lajolo, F. M., and Iaderoza, M.,** Fermentacao Semi-solid de raspa e folhas de mandioca, *Rev. Brasil. Mandioca,* 2(2), 77, 1983.

225. **Gregory, K. P., Reade, A. E., Santos-Minez. J., Alexander, J. C., Smith, R. E., and Mac Lean, S. J.** Further thermotolerant fungi for the conversion of cassava starch to protein, *Anim. Feed Sci. Technol.,* 2, 7, 1977.

226. **Alexander, J. C., Kuo, C. Y., and Gregory, K. P.,** 1979. Biological evaluations of two thermotolerant filamentous fungi as dietary protein sources for rats, *Nutr. Rep. Int.,* 20, 343, 1978.

227. **Toro, J. C.,** Alta proteina en yuca, in Curso Sobre Produccion de yuca, Medellin, Institute colombiano Agropecuario, Regional, 4, 189, 1975.

228. **Mikami, Y., Gregory, K. F., Levvadoux, W. L., Balagopalan, C., and Whitwill, S. T.,** Factors affecting yield and safety of protein production from cassava by *Cephalosporium eichhorniae, Appl. Environ. Microbiol.,* 43, 403, 1982.

229. **Khor, G. L., Alexander, J. C., Santos-Nunez, J., Reade, A. E., and Gregory, K. F.,** Nutritive value of thermotolerant fungi grown on cassava, *J. Inst. Can. Sci. Technol. Aliment,* 9, 139, 1976.

230. **Kuo, C. Y., Alexander, J. C., Lunsden, J. H., and Thomson, R. G.,** Subchronic toxicity test for two thermotolerant fungi as dietary protein sources for rats, *Nutr. Rep. Int.,* 20, 343, 1979.

231. **Balagopalan, C. and Padmaja, G.,** Fermentation of cassava starch for single cell protein production, presented at Nat. Symp. Production, Utilization Tuber Crops, Trivandrum, India 1985.

Chapter 8

CASSAVA FOODS

I. INTRODUCTION

Cassava is mainly utilized as human food in all the countries where it is cultivated, though it is gaining importance as a raw material for the industrial production of starch, sago, liquid glucose, and alcohol. Cassava constitutes the major staple food of more than 300 million people in the tropical countries and 70% of the total production is utilized for various food preparations.[1] Cassava roots are incorporated into the human diet in several forms. In many parts of the world it is either consumed directly after boiling, along with spices and salt, or mixed with several other vegetables. It is also subjected to various types of fermentation before being consumed.

Consumption patterns of cassava vary in different countries.[2] In Africa, nearly all cassava is used for human consumption. In Nigeria, gari, a fermented flour is marketed in urban areas and each year 120,000 tonnes of fresh cassava are sold in the Lagos market as gari. A good amount of the total production is utilized for human consumption in India, Indonesia, the Philippines, and Malaysia. In Central and South America, over 40% of the cassava production goes towards human consumption. A survey in Colombia showed that more than 90% of total production was sold off the farm. In Brazil, a large quantity of cassava is processed as either farina or flour.

Based on the FAO food balance sheets for 1975 to 77, estimates have been given on the cassava consumption pattern for different countries.[2,3] According to the balance sheet, from 1975 to 77, 500 million people in 24 countries consumed more than 100 cal/day in the form of cassava. In the tropical African countries, cassava provides an average of 230 calories per person per day. In Zaire and the Congo, the average intake is over 1000 cal/day. The average daily consumption in Indonesia is 2000 calories per adult. In Brazil and Paraguay more than 200 cal/day are obtained from cassava for an estimated population of 125 million people. In tropical South America as a whole, the average daily consumption is 150 to 160 calories per person.

II. COOKING QUALITY OF CASSAVA

Very wide variation in cooking quality among varieties has been observed. Some varieties consistently cook soft and mealy, whereas some cook well only under specific conditions (e.g., age of the crop, environmental factors etc.,) while a few others cook invariably hard and nonmealy. Reasons for such variation have been studied in detail.[3] Since it was necessary to specify some quantitative measure of the cooking quality, tests had to be developed to measure softness, mealiness, etc., under specified conditions and their gradings.

A physical measure of quality could be obtained by determining the weight and volume increase during cooking. Varieties which cook soft and mealy increase their weight and volume on cooking by about 5 to 10%, while the varieties with poor cooking quality lose weight and volume by over 30%. The external water used for cooking does not penetrate to the tubers during cooking. So the process of cooking cassava was identified as an expansion of starch granules by utilizing the water present within the tuber due to the heat supplied through the water surrounding the tubers.

The chemical changes that take place during cooking were also examined at different stages of cooking. The dry matter, sugar, starch, calcium, and phosphorus contents did not show any significant variation in most of the varieties. But disintegrated samples showed reduction in dry matter. Among chemical constituents, relation between softness and lower calcium content and higher P content was observed. Though higher starch content was invariably necessary for better cooking quality, not all varieties with high starch content showed good quality. Hence the role of distribution and properties of starch in determining the cooking quality was investigated.[3]

The starch and sugar contents in the outer region (just below the rind), the intermediate portion, and the innermost core portion of the tubers of many varieties were estimated at three regions: viz. proximal, middle, and distal. It was observed that varieties in which the starch distribution in the outer, intermediate, as well as the core regions was more uniform showed better cooking quality. On the other hand, the varieties with nonuniform distribution, especially with low starch content at the core, had poor cooking quality.

Better cooking quality is also coupled with low sugar content. In fact, the starch content at the core for poor quality tuber is lower than the sugar content at this region. It is expected that when the starch content is low at the core portion, the force of expansion cannot be sufficient to exert an outward pressure towards the intermediate region which has more starch. Thus, the core remains nonmealy while the intermediate region is partially mealy. This effect is confirmed by the fact that well cooked tubers expand in diameter whereas the diameter of poorly cooked ones decrease or remain same.

The swelling volume of starch extracted from different regions also varied significantly for the varieties which showed poor quality, while the values were more or less steady for better quality tubers. The wide range in swelling volume indicates low associative forces between the starch molecules and hence the tendency to break down on swelling. This leads to uneven swelling and cooking.

It was also observed that solubility of starch of all varieties depends on the amount of inherent water present for swelling.[3] When starch content is below a critical value of 20%, the solubility is higher than the solubility corresponding to starch contents above 20%. This means that free swelling, which leads to breakdown and higher solubility is inhibited in presence of reduced quantities of water available for swelling. This effect is very visible in regions of low starch content. Where the starch granules having suffered high breakdown, the area appears nonmealy.

Since other constituents may also be expected to play a role in quality, the effect of sucrose, glucose, and an electrolyte (sodium chloride) on swelling of starch was studied.[3] It was found that at increasing concentrations of sucrose, the solubility of starch increased considerably. This explains the cumulative effect of higher sugar content and lower starch content in imparting higher solubility of starch, which leads to a cohesive nonmealy nature.

In order to study the effect of other components on starch property, the extracts from different varieties were taken and starch was allowed to swell freely in these extracts. It was observed that extract of M4 invariably increased swelling volume of starch of all varieties, whereas other varieties showed variable effects. It was also observed that H97 extract at times had an exceptional property of bringing down the swelling volume of starches to very low levels. Experiments have been done to find out the role of the extract on the swelling of starch and hence on the cooking quality. A positive correlation has been reported between peak viscosity of cassava starch of different varieties and cooking quality.[5]

III. TRADITIONAL CASSAVA PREPARATIONS

A. Fresh Cassava Tubers as Food

The simplest and most popular mode of consumption of cassava tubers is in the form of mashed tubers. The fresh tubers after peeling of the skin and rind are cut into slices and boiled in water for 10 to 40 min until it is sold, depending on the variety. After cooking, the water is decanted and the boiled tubers are consumed with suitable dishes including those prepared out of fish or mutton. The boiled cassava tubers are also used in the preparation of some breakfast foods. The slices of fresh tubers in oil form a common snack in most of the households in Kerala, India. But in Srilanka, cassava roots are peeled, washed, cut into pieces, and boiled in water. The water is drained off and the pieces are eaten with curry. It is also cooked with spices and eaten. If bitter cassava is used, the unpleasant taste remains after cooking and the food is dangerous. Although boiling destroys the enzyme linamerase and drives off the hydrocyanic acid (HCN), the linamarin B itself is not destroyed and a long-term ingestion of it may lead to chronic cyanide toxicity in people whose diet lacks sufficient protein and iodine.

Roasting is another technique used for cooking the tubers. The fresh tubers are directly kept in country ovens and after sufficient periods of cooking in the fire (appearance of black charred skin is an indication of the completion of baking), the tubers are taken out and eaten.

In Brazil the tubers are often cooked in sugar syrup and eaten as a sweet meal. A soup or straw called Sacncocho or Cocido is obtained by boiling cassava roots with other vegetables.

B. Grating, Pounding, Baking, or Boiling

The techniques adopted in South America and southeast of the Amazon basin for the preparation of cassava are complex and have been described.[6] The cassava tubers are grated on prickly palm roots or spiny palm trunks and pounded into a pulp. The pulp is then squeezed by hand and cooked in one of a variety of ways. Several groups shape the pulp into pies or cakes which are then baked in hot ashes, that are sometimes being wrapped in a protective covering of leaves before baking. The pulp is also stored for use during times of food shortage. Balls of the pulp are sun dried, wrapped in leaves, and placed in baskets or buried in the ground. The cassava pulp is boiled, either by dropping pies or dumplings made from the pulp into boiling water or by stirring the pulp into water to form a sort of porridge.

1. Manicuera

In the northwestern Amazon region, boiled cassava juice is a slightly sweet drink popularly known as Manicuera. It is served late in the afternoon or early evening.

2. Mingao

This is another drink which is prepared by dissolving some of the fermented starch in boiling water and cooking it until thickened. It is a bland-tasting beverage, which is often flavored with palm fruits, pineapple, bananas, or lemon.

3. Fufu

The preparation of pastes by pounding is another method of consumption of cassava. Fufu is prepared by pounding peeled roots which have first been cooked by steaming or boiling. The sticky dough is then eaten with soups, particularly okra soup or stews of meat or fish.

4. Dumby

This is a Liberian food prepared from cassava.[6] After removing the skin and rind of cassava, the coarse central fibers are removed. The boiled tubers are placed in a wooden mortar and beaten with a heavy pestle. The beating requires considerable experience and skill. In order to prevent sticking, the pestle is dipped into water. As the mass becomes homogenous the pestle produces a loud crack. Pounding takes about 45 min as it is very hard. Dumby is normally eaten with a soup made from a variety of meat and vegetables. It is cut up into pieces, moistened with soup, and then swallowed whole.

5. Ampesi

The boiled roots may be eaten alone, mashed or whole or more usually accompanied by a sauce. They are often cooked with other foods such as beans or meat or cooked alone.

6. Farina

This is a granular product prepared in South America and in the West Indies. This is prepared from fresh tubers by peeling, grating, and squeezing out the juice in wooden screw press of basket work cylinder, forcing the pulp through a sieve, and finally roasting it slightly on a slow fire. Farina could be kept for several months and is consumed as a cereal usually in combination with other foods. In the Amazon region in Brazil it is prepared from yellow varieties of bitter cassava which have been soaked in water for 2 or 3 days, then peeled and grated. The resulting mash is mixed with grated fresh roots allowed to ferment for several days and preferably several weeks and then toasted.

7. Cassareep/Tucupay

The juice pressed out from the tubers during the preparation of farina is concentrated and various spices are added to obtain a type of sauce known as cassareep in the West Indies and tucupay in Brazil.[6]

8. Landang or Cassava Rice

In the Philippines, after squeezing out the juice, the pulp is made into pellets which are called landang or cassava rice. It is a popular article of food used as a substitute for rice and maize.

9. Macaroni

This is prepared by blending tapioca and ground nut flour with wheat semolina in the ratio of 60:15:25. It contains about 12% protein and promotes better growth than raw milled rice. An enriched macaroni containing 18 to 20 proteins and fortified with vitamins and minerals has also been developed for feeding weaned children and other vulnerable groups.

10. Cassava Pudding

In the Solomon Islands, the roots are grated and mixed with coconut or banana as a pudding.

C. Cassava Flour Preparations

The cassava tubers are boiled in water first and sun dried. The dried tubers are then pounded or ground into flour to be used for various preparations.

Cassava flour is used in a number of ways in South India and Southeast Asia. It can be made into a sort of porridge by mixing with water. Sometimes it is mixed with rice before cooking.

Traditional Indian foods such as chappathis, uppuma, puttu, iddlies, and dosa can be made from the cassava flour. In the Philippines, it is used in delicacies such as bibingka, seeman, and kalanay.

Cutlets are made by mincing the grated roots and mixing with fried onions, cashew nuts, black gram, and coriander leaves. The mixture is then made into balls dusted with maida flour and lightly fried. The flour can also be used for making breads, biscuits, salad dressings, custard powder, ice cream powder, flakes, vermicelli, etc.

Various delicacies have been suggested utilizing cassava flour. Some of the preparations suggested are listed here.

1. Cassava Dumplings

1 cup cassava flour
$^1/_2$ cup flour
$^1/_2$ teaspoon baking powder
$^1/_4$ cup milk
$^1/_4$ teaspoon of salt
1 tablespoon margarine
$^1/_4$ teaspoon sugar

Sift all dry ingredients into a bowl. Rub the fat and add milk gradually. Mix to a stiff dough. Shape into balls and cook for more than 20 min on the top of the soup.[7]

2. Cassava Fruit Cake

12 oz or 340 g cassava flour
6 oz or 170 g sugar
3 eggs
2 level teaspoons baking powder
6 oz margarine
12 oz mixed fruit
4 tablespoons milk

Grease and line a 7 in. cake tin. Place all ingredients in a large bowl. Beat together until all the ingredients are well blended and of a smooth pouring consistency. Place the mixture in the tin and smooth the top of the mixture. Bake in the middle of a moderate oven 450°F — remove and cool on a wire rack.[7]

3. Cassava Cakes

2 cups or 454 g cassava flour
2 teaspoons baking powder
2 tablespoons margarine
$^1/_3$ cup water and milk mixed
$1^1/_2$ oz or 14 g sugar
1 teaspoon salt
1 teaspoon vanilla

Sift the flour, salt, and baking powder. Dissolve the sugar in the water and milk and mix to a soft dough. Knead lightly, cut into small pieces. Roll into a ball, flatten to $^1/_4$ in. thick and fry in smoking hot oil until brown.[8]

4. Cassava Banana Fritters

2 tablespoons baking powder
1 tablespoon sugar
1 cup cooking oil
1 egg
1 cup sabah banana mashed
$^1/_2$ cup cassava flour
$^1/_2$ cut wheat flour
$^1/_2$ tea spoon salt
$^1/_2$ cup condensed milk

Sift and mix together the dry ingredients. Beat the egg and add to the milk. Add the rest of the ingredients and mix until blended. Fry by spoonfuls in deep hot fat until golden brown. Drain and sprinkle with sugar.[8]

5. Cassava Puttu

1³/₄ cup cassava flour
1 cup rice flour
¹/₄ cup coconut (Scraped)
salt (to taste)

Mix the flour well with required quantity of water and salt without lumps. Fill the mixture in the special container used for this preparation with the coconut scrapings in between. Steam for 10 min.[9]

6. Cassava Uppuma

500 g cassava
6 teaspoons fat
1 small piece ginger
2 teaspoons chopped coriander leaves
2 teaspoons bengal dhal
2 teaspoons black gram dhal
2 teaspoons mustard seeds
4 to 5 Nos Curry leaves
Salt (to taste)

Parboil cassava pieces. Chop green chillies and ginger finely. Fry mustard seeds, chopped ingredients, bengal gram dhal and black gram dhal in hot oil. Add curry leaves, water, and salt. When water begins to boil, add half-cooked cassava pieces. Stir and keep on a slow fire till cooked and dry.[9]

7. Cassava Masala Poori

300 g cassava
2 cups maida
¹/₂ teaspoon cumin seeds
1 onion (Big)
1 teaspoon Chopped coriander leaves
¹/₂ teaspoon green masala
Salt (to taste)
Oil for frying

Boil and mash the cassava well. Sieve the flour and add cumin seeds, masala powder, chopped coriander leaves, salt, and mashed cassava and knead well with a little oil. Divide into small portions and spread it into ¹/₂ in. thickness. Heat the oil and fry the poori till golden brown.[9]

8. Cassava Porotta

400 g cassava
1 cup wheat flour
4 green chillies
1 teaspoon ginger
1 teaspoon chopped coriander leaves
Salt (to taste)

Cook and mash the cassava. Grind the masalas and mix with the mashed cassava and divide into balls. Knead wheat flour with adequate water. Spread the wheat dough on the chappathi board. Keep the cassava masala mixture and cover with wheat dough. Flatten on the board. Cook on hot oiled tawa.[9]

9. Cassava Vattayappam

$1\frac{1}{2}$ cup cassava flour
1 cup milk
1 egg
$\frac{1}{2}$ cup sugar
2 to 3 drops essence

Beat the egg well. Add milk and sugar to the beaten egg and mix well. Mix the cassava flour with the egg milk mixture. Pour this on molds and steam.[9]

10. Cassava Idiappam

$1\frac{1}{2}$ cup cassava flour
salt (to taste)
$\frac{3}{4}$ cup coconut scrapings

Add salt and boiled water to the cassava flour and make into a soft dough. Use an idiappam mould and press out the dough. Spread the coconut scrapings over it. Steam it for 10 min.[9]

11. Cassava Iddli

$1\frac{1}{2}$ cup cassava flour
$\frac{1}{4}$ cup black gram dhal
Salt (to taste)

Soak black gram dhal and grind it. Mix the cassava flour with the ground dhal and salt. Make iddlies in the iddli steamer next day.

12. Cassava Dosai[9]

720 g cassava flour
1 cup black gram dhal
oil for frying
Salt (to taste)

Soak black gram dhal in water for 3 to 4 hr. Wash and grind the dhal into a fine paste. Mix cassava flour with water and make a pour batter. Add the ground black gram dhal paste to the cassava flour batter and make a thin batter. Add salt to taste and leave it overnight. Pour the batter on a hot griddle and spread it out with a ladle.[9]

D. Fermented Foods

Food fermentations are age old practices done under natural and controlled conditions. Fermentation processes usually improve the nutritional values of low protein and high carbohydrate foods. Fermentation can also impart characteristic flavor and aroma and also improve the palatability of food. Cassava forms a substrate for a wide variety of fermented foods in Africa, Asia, and Latin America. Solid state and liquid fermentations are the usual types of fermentation adapted before cassava is utilized for consumption and the procedure adopted in different countries differ widely. Most of the traditionally fermented foods are consumed immediately after fermentation, as it is produced for domestic uses. Only a very few items are produced on a commercial basis for later use.

1. Cassava Bread

Cassava bread is an important traditional fermented food of South America prepared by following the principles of solid state fermentation. Research and development activity for the preparation of bread based on cassava have been taken up by many international and national organizations throughout the world.

Dufour has described the traditional preparation of bread in the Northwestern Amazoan area by the Tukanoan Indians. This bread is called "Cazabe". The Tukanoan Indians use bitter cassava varieties for making bread. The freshly harvested roots are washed and then peeled. The tubers are then pulped normally by grating but some times by crushing in a mortar or between stones. Pressing separates the roots into liquids, starch, and fiber. This is followed by grating and straining. The resulting wet pulp is placed in a large basket strainer set on a tripod, rinsed with water, squeezed, kneaded, and pressed against the strainer to squeeze out the liquid.[10]

The extracted liquids which carry the starch suspension are collected under the strainer in a large clay pot. The starch is allowed to settle and the liquids decanted off the top to make juice.

Once separated, the starch and fiber are relatively stable and can be left in pots or a leaf-lined baskets for several days. Slight fermentation takes place under this condition and this fermented mash is used for bread making. In order to prevent further surface spoilage, the surplus mash is buried in leaf-lined pits and stored for longer periods.

Attempts have been made by several workers and organizations to improve the traditional bread preparations or develop composite flour technology for cassava based preparations by mixing other starches with cassava flour. Different wheat flours have been diluted with various proportions of cassava starch.[11] Mechanical dough development gave the best bread at 38°C and then added to the other constituents. The dough was mixed in a laboratory mixer at slow speed for 4 min. The water content was varied depending on moisture content of the cassava product used. The dough was scaled at 450 g, allowed to brew to a height of 11.5 cm, and baked 25 min at 218°C.

The minced fresh cassava scored the highest in physical assessment and it was prepared in a single operation and blended easily with dry ingredients used in bread making; the loaf of bread containing it also had the highest overall score (30) with a specific volume equaling 82.2% of the mean for control loaves. Blanched minced cassava was sticky, bulky, and difficult to blend with the dry ingredients. Preparations of acceptable breads using wheat flour supplemented with up to 20% flours and starches from cassava maize, rice, or yam has been also reported.[17]

Cassier et al.[18] have made good quality bread from pure and mixed flours of several tropical starchy crops like cassava, millet, sorghum, rice, and maize using 2 to 4% water in soluble purified and technical rye pentosan. With liquid dough loaves of starches and tropical crops, shelf life, texture, and crumb coherency were improved.

In the composite flour program development of bakery products and paste goods from cereal and noncereal flours, starches, and protein concentrates, the Food and Agriculture Organization (FAO) of the United Nations has given a description of two innovations being tested to employ larger quantities of cassava, corn, or sorghum flours in bread making: (1) Bread without gluten substances that agglutinate starch (Glyceral monosterate, 10% emulsion) are used to substitute wheat gluten. In order to compensate nutritional inferiority of all starch bread, considerable amounts of plant proteins are added in the form of peanut or soy flours producing a bread that is more nutritious than traditional ones. (2) The proportion of nonwheat flours could be increased to more than 10%.

Studies have been also made to improve the nutritional quality of cassava bread by incorporation of various other flours rich in proteins. Kim and De Ruiter[19] tested tuber flours including those derived from cassava, yarm, sago, and arrowroot for their suitability in bread making when combined with protein concentrates obtained from soybean, peanut, cottonseed, and fish meal. Cassava soy flour mixture, a corn-cassava-soy mixture gave good results. Kim used 80% cassava and 20% defatted soybean flour and found that addition of glyceryl monostearate (GMS) was required to improve the crumb structure of bread.[29]

Youngh et al.[21] studied the effect of adding (GMS), Polyoxyethylene stearate, and proteins (egg albumin and gliadin) in dough structure. A cassava starch dough (80% cassava plus 20% soybean flour or peanut flour) with 70 to 80% water and 1% GMS gave good results.

De Ruiter mixed cassava starch with soybean or ground nut flours using a GMS emulsion as an improving agent for bread making. The composite flour consisted of 70% wheat, 25% corn, cassava flour, or starch, and 5% soybean flour.[22]

While studying the rheological and baking properties of flour blends containing various levels of cassava starch, flour, or yam flour, Ciacco and Dappolonia[23] found that as the tuber flour level increased, developing time and stability of dough increased.

Rice, corn, and cassava starches were added to wheat flour to form three composite flours which yielded loaves of acceptability.[24] The rheological characteristics of the different flours showed a decrease in farinograph absorption from 680 to 63.2% and an increase in the mechanical tolerance index from zero to 10, 40, and 40 Brabender units (B.U.) for flours containing rice, corn, or cassava starch, respectively. The baked bread exhibited a range in loaf volume from 1688 to 1463 cc.

2. Gari

Gari is one of the most important items in the diet and forms a significant part in Ghana, Nigeria, Guinea, Benin, and Togo.

Cassava roots are washed, peeled, and grated and then the pulp is placed in cloth bags or sacks made from jute and left to ferment. The fermentation time varies from 3 to 10 days, according to taste. Various types of microorganisms are found during fermentation.

When fermentation is completed, the partially dried cassava pulp is taken out of the sacks and sieved to remove any fibrous material. It is then heated in wide, shallow iron pans and stirred continuously until it becomes light and crisp. Palm oil is added to prevent burning. Gari is a granular meal which is creamy white in color or yellow if palm oil is used in the cooking. Good quality gari is usually creamy yellow in color with uniformly sized grains and should swell to 3 times its volume when placed in water. For safe storage, moisture content must be below 12%.

Corynebacterium manihot and *Geotrichum candidum* are the two organisms which help in the fermentation of cassava in two stages. *C. manihot* breaks down cassava starch and produces various organic acids, including lactic acid, which lowers the pH of the mash and thus causes the hydrolysis of linamarin and lataustralin with the concomitant release of hydrogen cyanide. The acid pH of the cassava mash then provides favorable growth conditions for a second organism, *Geotrichum candidum*, which provides various aldehydes and esters that impart the characteristic taste and aroma of gari.[25] *Lactobacillus*, *Leuconostoc*, and *Alcaligenes* species are present in the fermenting cassava mash at a later stage.

Addition of exogenous linamarase preparation to fermenting grated cassava not only increased the rate and extent of detoxification but also consistently yielded gari with innocuous levels of cyanide.[26] A preliminary screening of several fungal isolates for their ability to synthesize linamarase resulted in the identification of two fungi *Penicillium steckii* and *Aspergillus sydowi*, capable of producing this enzyme in commercial quantities.

Gari is eaten plain, mixed with fresh coconut, or with roasted groundnut. It is also mixed into various dishes by adding hot or cold water and in combination with oils, spices, and soups. It is also mixed with wheat flour to make bread and pudding.

3. Meduame-M Bong

This is a fermented preparation in Cameroon. The roots are peeled, washed, cut into large pieces, and boiled for 30 min to 1 hr. The water is discarded and the roots are cut up further and these small pieces are then soaked in running water for 12 to 36 hr. The final product is called meduame-m bong and is eaten with sauces of meat, fish, peanuts, green leaves, etc.[6]

4. Attieke

This is a fermented produce in the Ivory Coast. The roots are peeled, steeped in water, and then ground into a paste and the paste is left to ferment for 2 days in jute sacks and pressed. Finally the paste is removed from the sacks, crumbled by hand, and steamed. The final produce has a sour taste and is eaten with milk or meat and vegetables.[6]

5. Chick Wangue

In this method, cassava is made into a paste after removing the rind and skin and soaking in water and pounding. The paste is sun dried or smoke dried. The wet paste is made into balls or chick wangue and are then placed on a screen over the hearth and normally left for about 15 days, although they can be left out longer. Fermentation is mainly by a bacterium and the species is not specified. After this, the leaves are removed and the black coating which forms during drying is scraped off. The dried paste is then ground and sieved to produce a flour. Sun-drying is achieved by spreading the loose wet paste onto mats and leaving it for 2 to 3 days. The paste is then crushed in a mortar or on a grindstone and sieved.

6. Kapok Pogari

This is a preparation of midwestern Nigeria. It is similar to gari preparation but the grated and fermented mass is not sieved before roasting, resulting in a produce with much larger particles than those of gari. Kapok Pogari is eaten dry with fish, ground nuts, coconut, or meat.

7. Peujeum

This is a traditional food of Java. The peeled roots are steamed until tender and allowed to cool and then dusted with finely powdered ragi (a mixture of flour and spices in which fungi and yeasts are active). The cassava mash mixed with ragi is wrapped in banana leaves in an earthenware pot and left for 1 to 2 days to ferment aerobically. The peujeum has a refreshing acidic and slightly alcoholic flavor and is eaten either as is or baked.[6]

8. Lafun

This is a fermented food of West Africa and the fermentation is mainly by a bacterium. Unpeeled roots are soaked in pots of water for 4 days. Afterwards, the peels are removed and the roots are sundried for 4 days.[6]

9. Wayana Cassava Cakes

The Wayana Indians in the rain forests of the Amazon region in South America extract the hydrocyanic acid from cassava by peeling and grating the root into mush, then squeezing the mush in a tubular wicker press hung from overhead beams. Once the poisonous juice has been extracted, the mush is turned into flour known as cassava and usually served as pancakes.

E. Fermented Drinks

1. Cassava Beer

In Uganda, cassava flour is mixed well with water and left to ferment for a week. It is then roasted over fire and put into a container to which water and yeast are added. After about a week the liquid is drained, sugar is added, and the beer is left to ferment for 4 days. It is drunk as is, or distilled into a more intoxicating drink called Wagari.

Becker has used cassava as a carbohydrate source as possible adjuncts for beer manufacturing.[27] Cassava was found satisfactory, provided the necessary sanitary regulations are kept. Cassava meal boiled for about 10 to 30 min gave the best results. It did not affect the taste of the beer, although a slight decrease in the hop action has occasionally been observed when the percentage of cassava was too high. Dhanuja and Singh[28] examined the possibility of the use of cassava starch for its suitability as a malt adjunct in brewing. The use of adjunct varied from 15 to 50%. Analysis of warts showed that reducing sugar content was in general comparable to the control wart. Alcohol content of the adjunct beer was slightly less than the control, and the adjunct beer was organoleptically comparable with the control. Rajagopal[29] has also attempted to produce beer utilizing cassava as an adjunct.

2. Beiju

Beiju is a naturally fermented cassava mass resembling Koji in oriental countries used for the production of an indigenous alcoholic beverage called liquira. The Microbial population in beiju was found to be predominantly molds and the number of colonies on PDA that developed from 1 g of beiju were 6×10^5 to 1.9×10^6. Predominant molds were *Aspergillus niger* and *Pacilomyces* sp. Moderate number of *Rhizopus* sp. resembling *R. delemer* and a small number of *Neurospora* were also found in beiju.[30]

3. Banu or Uala

This is a distilled liquor made from crushed and fermented cassava roots. Sometimes rice is also mixed with cereals to prepare beverages. Sun-dried cassava flour is crushed together with malted corn. Water is added to the mixture and the mash boiled for 12 hr after which time it is strained and cooled. This can either be drunk in this form as a sweet nonintoxicating drink or left to ferment into a sour alcoholic beer.

4. Kasili

The Wayana Indians boil cassava flour in river water and as the mash boils, the women chew cassava cakes and spit them into the pot so as to aid the fermentation process with saliva. The Wayanas are so fond of Kasili that they deliberately drink to it capacity, absorb the alcohol, then vomit and begin drinking again.

IV. PREPARATION OF NONCONVENTIONAL FOODS

A. Balanced Foods

The investigations carried out at the Central Technological Food Research Institute (CTFRI), Mysore and Food and Nutritional Laboratory, Hyderabad, India lead to the formulation of the following nonconventional foods:

1. Flours of cassava (45%), wheat (20%), groundnut (25%) and Bengal gram (10%).
2. Flours of cassava 70% and groundnut flour 30%. These food items produce a balanced ration consisting of 15% protein and 10% calories. These protein rich food formulae are being adopted to replace imported "CARE" food by indigenous formulations like "Balahar", "Kerala Indigenous Food (K.I.F.)". "Pushta atta" is another balanced food mixture of wheat flour, cassava flour, and groundnut flour.

B. Vegetable Cheese Process

The T.P.I. (Tropical Products Institute, London) vegetable cheese process consists of fermenting a cake of extruded cassava dough mixed with salts and inoculated with spores of *Rhizopus stolonifer*. The protein content could be increased from 0.6 to 4.3 on a dry weight basis. The process of vegetable cheese has been described in detail by Brook et al.[31]

C. Fortified Sago and Starch Products

A number of processes of fortification of sago and starch products, have been described.[32] Though sago is widely used in several parts of India for feeding invalids and infants, very little attempt has been made to elevate the nutritional status of sago which consists essentially of starch. The sago fortification project, of Salem, India, has evolved technology for producing fortified sago with a protein content of 14 to 16%, a protein efficiency ratio (PER) of 1.7 to 1.9, and is further fortified with vitamins and minerals. The unit has also achieved a fortification level of 1500 I.U. of vitamin A in the final product, which also contains appropriate levels of calcium, phosphorus, and iron. This food formulation satisfied all the requirements for a weaning food. But this technology cannot be adopted if sago is made under unhygienic conditions.

D. Enriched Noodles and Vermicelli

Enriched noodles and vermicelli containing cassava starch, wheat flour, soya flour, groundnut flour, vitamin A, calcium, phosphorus, and iron have been prepared by the sago fortification unit, Salem, India.

E. Nutritious Food Mixes

Nutritious food mixes have been developed utilizing cassava starch and mixing locally available cereals like bajra and maize. Further enrichment is done using groundnut flour. The cereals, cassava starch, and groundnut flour are roasted and then mixed with jaggery in equal proportions. The products have been found to have excellent taste and are readily acceptable to children.

F. Enriched Gold Finger

In the traditional process of manufacture of ordinary gold finger, 50% wheat flour and 50% cassava starch are used. Successful attempts have been made to substitute the 50% of wheat flour with soya flour and edible groundnut flour to make it more nutritious. This also eliminates large scale use of wheat flour in the manufacture of gold finger.

G. Cassava Rava

Cassava rava — a pregelatinized preparation modified and developed by CTCRI, India, could be prepared out of cassava as indicated in the process flow diagram (Figure 1).

Rava is a wheat-based convenience food used for the preparation of various breakfast recipes like uppuma and halwa. Conventionally, wheat semolina is used for this purpose. The properties of rava is based on its glutamin-gliadin content, which makes it swell to a small extent without breakdown. The method to be followed here is that the product should not become sticky, which requires controlled gelatinization of starch. The process for producing rava, consists of the following steps: (1) partial gelatinization of chips of cassava tubers, (2) drying them, and (3) pulverizing. By partial gelatinization, the granules swell to a small extent and give a granular form to the product. Care must be taken to avoid too much steaming or treatment in hot water, as this can lead to too much swelling, resulting in a cohesive texture on pulverization. Hence, the steam treatment has to be carried out only to a minimum extent. It has been found that a steam treatment of less than 5 min, at 5 psi of steam, or immersion in boiling water for less than 5 min are ideal for producing good quality product. The moisture content at this stage increases by 10 to 15% more than the original moisture content in the tubers.

After draining the water, chips are spread out on mats in the sun or in a mechanical dryer (temp. 70°C). The moisture content is brought down to around 15%. At this stage, the chips are hard.

The dried chips are then pulverized in a hammermill, to a fraction between a granule size of 0.5 to 3 mm, and a sieve size of 20 to 30 mesh.

Rava could be utilized for the preparation of traditional South Indian preparations, uppuma, and sweets.

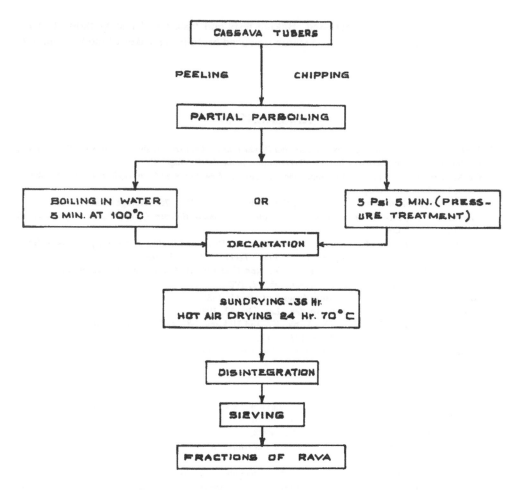

FIGURE 1. Process diagram for production of Rava.

FLOW DIAGRAM FOR GARI PREPARATION

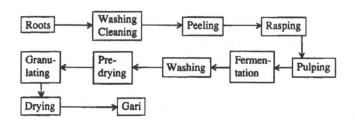

H. Putto

Putto is the most popular fermented rice preparation in Philippines similar to iddli in India. Successful attempts were made to substitute rice flour with cassava flour in putto preparation and that at a 50% substitution, the volume of putto was nearly the same as that made with rice flour.[33]

I. Biscuits and Cakes

Strong locally made wheat flour could be softened by blending with cassava starch for use in biscuits and cake manufacture. The effect of starch at levels of 10, 25, and 50% (flour unit) on dough characteristics showed that the strength of wheat flour decreased as the levels of cassava starch increased. Baking tests and taste panel results showed that a flour mixture containing 10% cassava starch produced biscuits and cakes of the best quality.

REFERENCES

1. **Nair, R. G.,** Utilisation of cassava, in *Cassava Production Technology,* Hrishi, N. and Nair, R. G., Eds., Central Tuber Crops Research Institute, Trivandrum, India, 1978, chap. 10.
2. **Cock, J. H.,** Cassava plant and its importance, in *Cassava New Potential for a Neglected Crop,* Westview Press, London, 1985, chap. 1.
3. Food and Agriculture Organization (FAO), Food balance sheets, 1975-76, Average, Rome, 1980.
4. **Moorthy, S. N.,** *Cassava Starch and its Modifications, Technical Bulletin,* Central Tuber Crops Research Institute, Trivandrum, India, 1985, 32.
5. **Olorunde, A. O., Aworh, O. C., and Numfor, F. A.,** Predicting quality of cassava products with the aid of chemical and rheological properties, *J. Food. Technol.,* 16, 447, 1981.
6. **Lancaster, P. N., Ingram, J. S., Lin, H. Y., and Coursey, D. G.,** Traditional cassava based foods survey of processing techniques, *Econ. Bot.,* 36, 12, 1982.
7. **Anon.,** Recipes with cassava, *Cassava News Lett.,* 9, 11, 1985.
8. **Anon.,** Recipes with cassava, *Cassava News Lett.,* 10, 12, 1986.
9. **Prema, L., Vimala Kumari, N. K., Usha, V., and Ukkuru, M.,** *Indian Dishes with Cassava,* Kerala Agricultural University, Trichur, India, 1982, 147.
10. **Dufour, D. L.,** The uses of bitter cassava in Northwestern Amazon, *Cassava News Lett.,* 8, 6, 1984.
11. **Dendy, D. A. V., James, A. W., and Clark, P. A.,** Working of Tropical Products Institute on the use of non-wheat flours in bread making, *Proc. Symp. on the Use of Non-wheat Flour in Bread and Baked Goods Manufacture, London,* 1971, 1.
12. **James, A. W.,** Bread making with cassava starch and coconut protein isolate, in *Development of a Wet Coconut Process Designed to Extract Protein and Oil from Coconut,* Dendy, D. A. V., and Timmins, W. H., Eds., Tropical Products Institute, London, 1973, 29.
13. **NG., E. G. and Khor, G. L.,** Use of cassava and winged bean flour in bread making, *Pertanika,* 1, 7, 1978.
14. **Rasper, V., Rasper, J., and Mabey, G. L.,** Functional properties of non-wheat flour substitutes in composite flours. 1. The effect of non-wheat starches in composite doughs, *Canadian Institute of Food Science and Technology Journal,* 7, 86, 1974.
15. **Knight, J. W.,** Speciality food starches, in cassava processing and storage, *Proc. Interdisciplinary Workshop,* Thailand, Araullo, E. V., Nestlle, B., and Cambell, M., Eds., IDRC, Ottawa, 1974, 77.
16. **Crabtree, J., Kramer, E. C., and Baldrj, J.,** Cassava harvesting and processing, Proc. Workshop, Cali, India, 1978, 24, 49.
17. **Tsen, C. C.,** Using non-wheat flours and starches from tropical crops as bread supplements, *Proc. International Conference on Tropical foods, Chemistry and Nutrition,* Inglett, G. E., and Charalambous, G., Eds., Academic Press, New York, 1979, 239.
18. **Casier, J. P. J., Paspe, G. M. J. D. E., Williams, H. E., Goffings, G. J. G., Hermans, J. L., and Noppen H. E.,** Bread production from pure flours of tropical starchy crops. III. From pure and mixed flours of cassava, millet, sorghum, corn, rice and the starches, in *Tropical Foods, Chemistry and Nutrition,* Ingold, G. E. and Charlambous, G., Eds., Academic Press, New York, 1979, 279.
19. **Kim, J. C. and De-Ruiter, D.,** Bakery products with non-wheat flours, *Baker's Dig.,* 43, 58, 1961.
20. **Kim, J. C.,** Bakprodukten wit meel vantrope Scheland, bouwage wasen, *Voedingsmiddelen Technol.,* 2, 1, 1971.
21. **Yongh, G., Slim., T., and Greve, H.,** Bread without gluten, *Baker's Dig.,* 42, 24, 1967.
22. **De-Ruiter, D.,** Bread making with non-wheat flour, Proc. Symp. on the use of Non-wheat Flour in Bread and Baked Foods Manufacture, London, 1970, 7.
23. **Ciacco, C. F. and Dappolonia, B. L.,** Baking Studies with cassava and yam. II. Rheological and baking studies of tuber-wheat flour blends, *Cereal Chem.,* 59, 4, 1982.
24. **Seyam, A. M. and Kidman, F. C.,** Starches of non-wheat origin, their effect in bread quality, *Baker's Dig.,* 49, 25, 1975.
25. **Akinrele, I. A.,** Fermentation of cassava, *J. Sci. Food Agric.,* 15, 589, 1964.
26. **Ikediobi, C. O. and Onyike, E.,** The use of linamarase in gari production, *Proc. Biochem.,* 17, 2, 1982.
27. **Becker, K.,** Non-grains adjuncts, *Brewer's Dig.,* 1946, 45, 1946.
28. **Dhaniya, S. S. and Singh, D. P.,** Adjuncts in brewing. II. Tapioca starch, *J. Food Sci. Technol.,* 16, 146, 1979.

29. **Rajagopal. M. V.,** Production of beer from cassava, *J. Food Sci.,* 44, 532, 1977.
30. **Park, Y. K., Zenin, C. T., Veda, S., Martins, C. O., and Martins neto, J. P.,** Microflora and Beiju and their biochemical characteristics, *J. Ferment. Technol.,* 6, 1, 1982.
31. **Brook, E. J., Stanton, W. R., and Bridge, A. M.,** Fermentation methods for protein enrichment of cassava, *Biotechnol. Bioeng.,* 11, 1271, 1969.
32. **Takkar, P. K.,** Supplementary value of groundnut flour blends and skim milk powder to a maize, tapioca diet, *Food Sci.,* 11, 181, 1982.
33. **Yoshi, H. and Garcia, V. V.,** Processing of cassava into fermented foods, in *Tropical Root Crops, Postharvest Physiology and Processing,* Uritanl, I. and Reyes, Eds., Japan Scientific Societies Press, Tokyo, 1984, 205.

Chapter 9

CASSAVA STARCH

I. INTRODUCTION

Starch is widely distributed throughout the plant kingdom and may be considered a counterpart of glycogen in the animal kingdom. It exists as the major reserve carbohydrate of most higher plants, especially in the tuber crops. The starch in tuber crops exists relatively free from lipids and proteins, and hence its extraction and purification are relatively simple. The high starch content from cassava, which can be grown under low management conditions, makes it an important industrial crop in addition to being a calorie-rich food crop.

Starch also exists in transient form in the leaf tissues, where it is synthesized in the presence of light and broken down during darkness. This takes place in the chloroplasts, whereas in the amyloplasts, the starch granules are continually deposited as reserve food.

II. CHEMICAL STRUCTURE OF STARCH

Chemically, starch is a polymer of glucose units joined by α-linkages compared to cellulose which has β-linkages. The α-linkages, being less stable compared to β-linkages, render starch relatively labile. Before the 1940s, starch was considered to be a single polysaccharide having a complex molecular structure, but Schoch and Meyer[1,2] proved the existence of a mixture of two polysaccharides. Both are polymers of α-D-glucopyranose units, the major component-amylopectin-having a branched structure, while amylose, the minor component, has a linear structure. Amylopectin has, in addition to α(1-4) linkages found in amylose, α(1-6) D-glucoside linkages, which contribute to difference in various properties between the two components. Amylose has been found to have a DP of several thousand glucose residues (mol wt of 1.5 to 10 × 10⁴); it has a high affinity for iodine, producing a dark blue color. The color is due to formation of a complex in which the iodide ions fit into the helical structure assumed by amylose in solution and forms the basis for determination of amylose content in starches. Most plant starches contain approximately 15 to 25% amylose. In cassava, the value is normally 16 to 18%,[3] though values as high as 25% have been reported with some varieties.[4] It has also been established that amylose is present both in crystalline and amorphous forms. The latter is easily leached out by water and hence is also called soluble amylose. The soluble amylose content is found to be approximately 50 to 75% of the total amylose content in cassava starch.[5] Soluble amylose in starch has been considered responsible for the stickiness of some rice varieties, though such correlation could not be found for cassava. The tendency of linear amylose molecules to align themselves parallel one another forming hydrogen bonds between them leads to spontaneous precipitation in aqueous solution and the phenomenon is known as "retrogradation". This has been found to be particularly prominent for high amylose starches and cereal starches, whereas it is less pronounced for tuber starches. In addition, the tendency to retrograde can be reduced by derivatization, which prevents parallel association through hydrogen bonds.

β-Amylase degrades amylose to maltose to an extent of 70 to 95% by cleavage of alternate α(1-4) linkages. The incomplete conversion of amylose by amylase has led to the theory that amylose contains occasional branching.[6]

The molecular weight of amylopectin (10⁷) is higher compared to amylose and it is made up of chains containing approximately 20 to 25 α(1-4) linked glucose residues which are interlinked to form a branched chain structure. It contains 94 to 96% (1-4) and 4 to 6% (1-6) linkages. The structure of amylopectin has drawn the attention of scientists. Howarth et al.[7] suggested a laminated structure. Staudinger and Husemann[8] suggested a herring-bone structure. Another structure proposed by Meyer and Bernfield[9] is a randomly branched model. New biochemical and chromatographic techniques helped Gunja-Smith to modify Meyer's structure.[10] Robin and co-workers[11] suggested a Nikuni ''cluster'' structure. In this model, A and B chains are linear and have a $\overline{DP}$ of 15 and 45, respectively. B chains are the backbone of the amylopectin molecule and extend over 2 or more clusters. Each cluster consists of 2 or more closely associated A chains. The associated regions of A chains form the crystalline region within the molecule. The amorphous regions occuring at 60 to 70 Å intervals contain the majority of α(1-6) linkages and are relatively susceptible to hydrolytic agents. Overall, the amylopectin molecule is 100 to 150 Å in diameter and 1200 to 4000 Å long. In the granule, amylose may be located between the amylopectin molecules.

The molecular weight of amylopectin from different starches is almost the same by light scattering techniques (4 to 5 $\times$ 10^6).[12] Using unmodified and modified cassava starch amylopectin and pullulanase, and β-amylase debranching in forward and reverse sequence, followed by gel permeation chromatography on a Sephadex® column, Hood and Mercier[13] not only established the structure of amylopectin, but also the positions at which substitution occurs. The amylopectin of unmodified cassava starch has a molar ratio of $\overline{DP}$ 45: $\overline{DP}$ 15 chains of 7.5:1. Most of the modifying groups are located in the amorphous regions, rich in (1-6) linkages.

The branched nature of amylopectin leads to properties like higher stability in solution. Iodine gives a purple color with amylopectin. β-Amylase only partly hydrolyses the molecules at the exterior chains. This has an important application in determining the exterior chain length of amylopectin (glucose units removed + 2). The majority of amylopectin molecules have an external chain length of 10 to 18 and an interior chain length of 5 to 9.

Although the molecular structure of starch has been unravelled to a great extent, knowledge of how these molecules are arranged within the granules is still limited. Recently, French[14] has summarized the theories. X-ray diffraction patterns have been of help to find the crystalline nature of starch. All common starches except high amylose corn starch possess well defined X-ray patterns. Cereal starches give an A pattern. Cassava and sweet potato possess an A pattern, while Dioscorea sp. show an A/B pattern. Potato gives a B pattern. These patterns differ mainly on the intermolecular space between the various helices made up of amylose/amylopectin outer chains. The A pattern is more compact. An intermediate C pattern has been observed for some starches while gelatinized starch gives a V pattern for all starches. The crystalline nature indicates a definite pattern of arrangement of amylopectin, while amylose has only a small role. Based on enzymatic and freeze etching studies, various models have been suggested. Finkelstein and Sarko[15] interpreted light scattering data to suggest that layers differ between cassava and potato starches. According to Yamaguchi et al.[16] the rings occur at irregular 1200 to 1400 Å intervals and probably represent the overall length of individual amylopectin molecules. They suggest that a single amylopectin molecule commences at one growth ring and terminates at the next. Each molecule is made up of many 70 Å clusters. The presence of lamellar structures supports the theory that starch granules grow during biosynthesis by apposition to produce an ''onion-like'' layered structure. The molecular structure of starch is related to the crystalline properties of the granule. Starch molecules are oriented radially within the granules. Pairs of outer $\overline{DP}$ 15 chains of amylopectin can form double helices which contributes to X-ray patterns.[17] Amylose molecules may be randomly spaced between the amylopectin molecules; some, which are oriented with amylopectin/amylose chains, being crystalline, while others form the amorphous region (Figure 1).

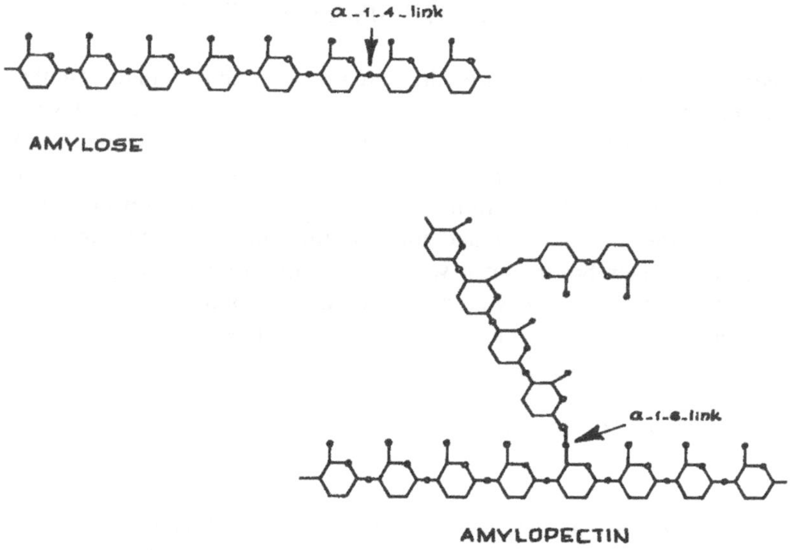

FIGURE 1. Structure of amylose and amylopectin.

III. ENZYMES IN SYNTHESIS AND HYDROLYSIS OF STARCH

The enzyme phosphorylase was the first one recognized in the synthesis of starch and the reaction is

$$(G)_x + \text{glucose} - \text{1-phosphate} \rightarrow G(1{-}4){-}(G)_x + Pi$$

The enzyme has a molecular weight of 207,000. However, the presence of another enzyme and also recognition of the catabolic role of phosphorylase led to reduced importance of this enzyme in starch synthesis.

The role of starch synthase was first highlighted by Leloir et al.[18] and the enzyme catalyzed the transfer of glucose units from UDPG to the starch granules and, in particular, to the nonreducing end of certain amylose and amylopectin molecules. Later on, ADPG was found to be a better donor. The reaction may be given as

$$\text{ADPG} + (G)_y \rightarrow G(1{-}4){-}(G)_y + \text{ADP}$$

However, the enzyme is difficult to be isolated pure and many points regarding its activity have still not been clarified.

Both the enzymes give only chain lengthening reactions and hence the necessity of a primer (at least malto-tetraose) exists. But later on, the presence of soluble enzymes which synthesize polysaccharides in the absence of a primer has been reported. The polysaccharides themselves may act as primers for action of phosphorylase and starch synthase.

Thus starch synthesis occurs in a series of reactions of different enzymes. The sequence — initially phosphorylase, followed by UDPG-starch synthase, and finally by ADPG-starch synthase — has how been recognized.

The enzyme responsible for the synthesis of branched amylopectin fractions has also been studied. The enzyme — Q enzyme — produces $\alpha(1\text{-}6)$ D-glucosidic linkages by action on (1-4) linked glucose residues. The reaction involves transfer of a short chain of glucose residues from $\alpha(1\text{-}4)$ linkage to $\alpha(1\text{-}6)$ linkage. The process appears to be irreversible. The amylopectin is formed by combined action of chain lengthening enzymes as well as Q enzymes. There is evidence that linear chains of $\overline{DP}$ 50 are required to act as an efficient donor or acceptor substrates.

IV. BIOSYNTHESIS OF STARCH

It has been observed that during growth, starch content and average granule size of the two components increase and the granules grow by apposition, in successive and gradual addition, of new polysaccharides from the outside. Though the basic synthesis may be similar for different plants and tissues, the relative roles of enzymes may be different. In addition, the enzymes may occur in different forms, eg., soluble and bound forms of starch synthase. It was suggested earlier that amylose and amylopectin synthesizing systems were separated by a semi-permeable membrane, but it was later modified that amylopectin itself may be acting as a steric barrier to Q enzyme acting on amylose molecules. It is not clear even now, why some (1-4) D-glucose molecule chains are subject to chain lengthening, while others are converted to amylopectin.

V. ENZYMATIC DEGRADATION OF STARCH

A. Phosphorylase

This enzyme, in presence of inorganic Phosphate, degrades (1-4) linked D-glucose residues with formation of G-I-P. Although phosphorylase can degrade amylose and outer chains of amylopectin, there is no evidence that it can degrade starch granules without initial partial hydrolysis by α-amylase.

B. α-Amylase

It is widely distributed in most of the plants and many animal and microbial sources. It acts by random hydrolysis of amylose and amylopectin, as indicated by rapid decrease in iodine affinity and viscosity, but with only a limited production of reducing sugars. However, it cannot hydrolyze α(1-4) D-glucoside linkages adjacent to nonreducing end groups or to α(1-6) D-glucosidic inter-chain linkages. With amylose, the products are maltose (90%) and either maltotriose or glucose. With amylopectin, linear segments are hydrolysed leaving behind a branched oligosaccharide having a $\overline{DP}$ 5 to 10 (α- dextrins).

C. β-Amylase

This occurs only in certain plants, noticeably sweet potato. It catalyzes degradation of starch type polysaccharides step-wise producing maltose. The β-amylases attack the glucosidic bond of penultimate nonreducing end of starch, but is blocked by α D- (1-6) branch linkage. Hence only 55% of amylopectin is converted to maltose leaving 45% high molecular weight limit dextrin.

D. Amyloglucosidase

Amyloglucosidase, known also as α-amylase, hydrolyses, stepwise, successive (1-4) D-glucose in the α-form. They can also hydrolyse (1-6) D-glucosidic linkages at a slower rate. They do not, however, occur in the plant kingdom, but in many fungi.

In addition, there are some other enzymes which are also involved in hydrolysis of starch, viz. glucosidase, R-enzyme, pullulunase, dextrinase, etc.

In vivo degradation of starch may be mainly brought about by α-amylase action which releases both reducing sugars and intermediate dextrins which are further degraded by enzymes such as dextrinase, β-amylase, amyloglucosidases, etc.

VI. PROPERTIES OF CASSAVA STARCH

A. Color and Appearance

Cassava starch, when processed properly, is very white in color. When the tubers are crushed without removing the rind, there is a dullness in the color. The reduction in whiteness not only affects the quality but also the price. In sago manufacture, the product obtained by using the starch extracted after removing the rind possess a greater whiteness, and demands a higher price compared to the one made without removing the rind. Similarly, for textile industries, the starch possessing maximum whiteness demands a higher price.

The method usually used for determination of color is very arbitrary, viz. visual. The visual method involves placing a small pile of starch on a porcelain plate and placing adjacent to it, similar piles of starches which have been kept as color standards for matching.[19] Using a clean, clear glass plate, the piles are carefully flattened so that they touch each other to form a continuous layer. The sample and standard are compared in diffused daylight, by noting the absence or presence of a boundary between them.

A colorimetric method involves comparison of the absorbance of solution of the starch to a standard. About 10 g of starch is suspended in 2% alcoholic sodium hydroxide (200 mℓ), boiled, cooled for 30 min with continuous stirring, filtered, and the absorbance of the solution is noted using a colorimeter.[19] However, this method is defective in that it assumes that the color, if any, is due to alcohol soluble impurities. A more reliable method is to use a Lovibond tintometer. The color reflected from a smooth surface of powdered starch is matched by interposing standard Lovibond series 200 red and series 510 yellow color glasses in the light reflected from the smooth white surface of a magnesia block placed beside the sample.

The Corn Industries Research Foundation has prescribed a standard procedure which specifies the spectro-photometer as a Beckman Model B instrument, or an equivalent equipped with the integrating sphere diffuse reflectant attachment with beam-expanding lenses and a blue sensitive phototube, the reflectant standard, vitrolite glass block secondary standard, and calibrated against freshly prepared magnesium oxide.

Processing or drying, especially in the open, brings in a lot of dust and foreign matter. One method used to remove or estimate the foreign matter involves sieving a definite weight of the dry powdered starch through a 200 mesh screen or through No. 17 bolting silk until the starch leaves the foreign matter on the screen. Instead of the dry starch, a suspension of 50 g of starch in 500 mℓ distilled water may also be used. To measure specks, the starch is spread out on paper, a sheet of glass with a 100 mm square marked on it, is placed flat on top and the number of specks enclosed in the square is counted. Suspension of the weighed quantity of starch in water or an organic solvent having specific gravity about 1.5 may also be used to separate and estimate the foreign matter. Improper storage can lead to moldiness which will contribute to bad odor. Naturally such an odor is not desirable. The most common organic acids produced are lactic and butyric acids. For testing the presence of any odor, water at 85 to 90°C is added to starch kept in a beaker, stirred and the odor noted. The odor can also be detected by enclosing the starch in a tight-fitting jar, heating at 80° (for 16 hr, and noting the presence of any odor upon opening).

B. Hydrogen Ion Activity or pH

The normal pH of cassava tuber is 6.3 to 6.5 and the starch slurry also exhibits a similar value. The specifications vary for the pH. The Indian Standard Institution (ISI) has allowed a pH range of 4.7 to 7.0 for edible starch.[20] But the U.S. standards by the Tapioca Institute are much more stringent and consider the best grade starch as that with a value of 4.5 to 6.5 and with low acidity.[21] The pH and acidity of the different grades are given below.

Grade	Initial pH	mℓ of N/10 Acid required to reach pH 3
A	4.5—6.5	2.5
B	4.5—6.5	6.0
C	4.5—6.5	Titration not required

Determination of acidity may be carried out either by titrimetry or electrometry. Acidity may also be categorized as extractable acidity or paste acidity. In titrimetry, 10 g starch is stirred in 100 mℓ distilled water for 30 min, gravity filtered, and titrated with 0.1 N sodium hydroxide using phenolphthalein as indicator, and it gives the extractable acidity. For determination of paste acidity, the starch (10 g) is dispersed in 300 mℓ water, the suspension is boiled for 10 min with occasional shaking and titrated immediately, and the value is expressed as milliequivalents of acid per unit sample weight.

In determination of pH by using glass electrode, continuous stirring is required to prevent starch from sticking to the electrode and giving erratic results.[22]

C. Flow Properties

The readiness with which powdered starch flows can be measured shaking dry starch in a covered sieve and noting the amount of starch passing through the sieve in a definite time.[23] A more simple test involves placing a small sample in a conical shape on a horizontal plate which can be inclined and is provided with a scale behind it showing the angle of inclination. The plate is slowly inclined at a constant rate and the angle at which the starch starts moving down is noted. A more mobile starch moves at a smaller angle. Cassava starch, perfectly dry, has a good mobility.

D. Size of Granules

Cassava starch granules size as determined microscopically gives a value of 5 to 40 μm: with larger ones in the range of 25 to 35 μm and the smaller ones in the range of 5 to 15 μm. Using a Coulton counter, the value has been determined to be lying between 6 to 35 μm.[4,24-26]

E. Shape

Under a light microscope, cassava starch granules exhibit varying shapes, but mostly are round with a truncated end.[27-29] Scanning electron microscopy has shown that granules are round in shape with a flat surface on one side containing a conical pit with a small projection at the center. A well-defined excentric hilum is observed. In polarized light, crosses of moderate to strong intensity are visible. Scanning electron microscopy has also shown that granules possess a smooth outer surface (Figure 2).[30] Freeze etching studies have indicated that the organization within the granules is not homogenous.[31] Particles were observed on the granule fracture faces and these particles had a size of 4 to 10 μm. In contrast to the fracture faces, the outer surface of granules were smooth and particles were found to be absent.

F. Fine Structure

The fine structure of all starches including cassava starch has been the subject of many studies. But, it is still not very clear. Physical and enzymatic methods have been used for the studies. The X-ray diffraction pattern of cassava starch has been found to be A with characteristic peaks at 12, 9, and 8°30'.[32,33] Earlier, many workers had reported that cassava starch belongs to C pattern. An A pattern, observed for most cereals, indicates a closer and more organized structure and occupies only a fourth of the volume occupied by B pattern starch, e.g., potato.[17] Since originally it was thought that only amylose is responsible for the crystalline nature of starch observed by X-ray diffraction, the X-ray results were interpreted as indicating the conformation of the amylose molecules, and they were assumed to exist in parallel standed right handed double helices and in an antiparallel fashion with an orthorhombic unit cell structure.[17] However, studies using enzymes have indicated that even amylopectin molecules can contribute to the crystallinity; in a thorough and detailed study using pullulanase and amylase, amylopectin structure has also been proposed.

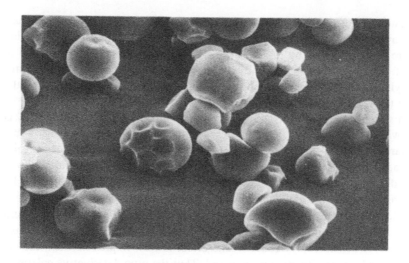

FIGURE 2. SEM photograph of starch.

G. Molecular Weight

The relatively higher amylopectin content of cassava starch can be expected to contribute to higher molecular weight and has been found to be experimentally so. Potentiometrically determined, the average molecular weight of cassava starch has a value of 215,000 g/mol[2] compared to 30,500 for amylomaize; 130,000 for wheat starch, 224,500 for potato starch, and 276,000 for waxy, maize starch.[34] These values have been further confirmed by low alkali numbers and reducing values (ferricyanide number) for cassava starch. Thus, Schoch and Jensen[35] found the following values for alkali numbers:cassava 5.9 to 6.9; maize 9.8 to 12.1; wheat 9.7 to 11.5, and potato 5.7 to 6.9.

However, the results on isolated amylopectin do not show such vast differences between the different starches.[12]

H. Amylose Content

Cassava starch has been reported to have an amylose content between 8 to 29%, but generally lies in the range of 16 to 18%.[3] The starch expresses the properties similar to all amylopectin starch, e.g., high viscosity, low retrogradation tendency, and good sol stability.[36] In fact, it is these properties which makes it very suitable for many food applications.

There has been a general tendency to associate cohesiveness of cooked tubers with higher amylopectin contents. However, recent results have indicated that the soluble amylose rather than amylopectin contributes to the cohesiveness of cooked tubers.[5] The soluble amylose content in cassava is of the order of 10 to 12%, indicating that a major portion of the amylose exists in the noncrystalline regions.

I. Gelatinization Temperature

Starch granules are made of amylose and amylopectin molecules associated inter- and intramolecularly by hydrogen bonding, either directly or through water hydrate bridges, to form micellar regions. The strength of the micellar networks decides many of the properties of the starch granules. When an aqueous suspension of starch is heated, the water molecules around the granules disrupt the hydrogen bonding, enter the granules, and the granules swell. This hydration-cum-swelling is an irreversible process and is termed "gelatinization". The gelatinization process does not occur abruptly, but after a critical temperature is reached. At this point, some granules swell tangentially, and they lose their polarization crosses and also their birefringence. Both these actions originate at the hilum and move to the periphery of the granules. Since the degree of association varies between granules, the granules gelatinize over a range of temperature, around 10°C. It has been found that gelatinization temperature is not affected by the presence of linear components since normal and waxy starches of the same species gelatinize over the same temperature range. Three factors accompanying gelatinization are (1) loss of birefringence, (2) increase in optical transmittancy, and (3) rise in viscosity.

Cassava starch possesses a gelatinization temperature range of 58.5 to 70°C, compared to potato which has a value of 56 to 66°C and maize 62 to 72°C.[36] The lower value for cassava starch compared to maize and the higher value compared to potato again indicate that associative forces in cassava starch lie in between those of maize and potato. Thus any treatment affecting the associative forces will affect gelatinization temperature. By derivatizing or cross-linking, the associative forces between hydroxyl groups can be expected to be reduced, leading to lower gelatinization temperature. This has been experimentally proven,[37,38] e.g., by forming the various esters, whereby the gelatinization temperature of cassava starch is considerably reduced. The effect is further proven when the size of the organic acid used for esterification is increased. Thus the propyl derivative gelatinizes earlier than the acetyl derivative.[37]

The effect of the addition of various reagents which combine with the starch molecules, can also be noticed on the gelatinization temperature. It has been noticed that the gelatinization temperature is drastically affected by the addition of different surfactants.[39,40] The values also depend on the hydrophilic-hydrophobic nature of the surfactant and size of the hydrophilic portion of the surfactant. A surfactant like potassium palmitate or potassium stearate, which can fit very closely into the starch helix, preventing easy penetration of water molecules, increases the gelatinization temperature considerably. In contrast, the surfactant cetyl trimethyl ammonium bromide, with a very bulky hydrophilic head, increases the gelatinization temperature to a much lower extent. Similarly, sodium lauryl sulfate, which has a 12-carbon hydrophobic portion, is less easily bound to the starch helix, and, hence, the increase in gelatinization temperature is much less. These results further indicate the helical nature of starch molecules in the solid form and also the micellar nature of starch in its suspension in water.

The solvent used to gelatinize starch also plays a role in determining the gelatinization temperature.[41-43] When different solvents were used, the gelatinization temperature varied with the size of the solvent molecule as well as the nature of the solvent. Glycerol required a much higher temperature to gelatinize starch in view of its much bigger size, and hence reflects the difficulty of the solvent to penetrate the starch helix and gelatinize the starch. Presence of hydrophobic portions also prevents easy gelatinization of starch. The gelatinization temperature also depends on the type of pasting of starch granules. It has been observed that wherever there is a single stage gelatinization, the range is much smaller compared to two-stage gelatinization. The effect of varietial differences on the gelatinization temperature is also evident. In a study of the cooking quality of cassava in relation to the gelatinization temperature of starches of different varieties, Moorthy and Maini found that the varieties which show lowest gelatinization temperature have poorer cooking quality, probably because easier gelatinization leads to easy breakdown of starch.

J. Viscosity

Viscosity is another important property of starch solutions, which makes it useful in many industries, e.g., the role of starch as a thickener in food industries, and as a sizing and finishing agent in textile and paper industries. When an aqueous concentrated suspension of starch is heated above its gelatinization temperature, the individual granules gelatinize and swell rapidly and freely until they consume almost all the available water. As a result of the high swelling, the granules become increasingly susceptible to disintegration by mechanical shear. In the initial phase, the starch solubles leach out of the granules and diffuse into the adjoining aqueous phase.[36] However, when the whole volume is occupied by the swollen granules, part of the solubles diffuse back into the swollen granules from the solution, until an equilibrium is reached. Thus, the apparent viscosity of a starch solution is caused not only by individual swollen granules, but also by presence of starch solubles which hold the swollen granules together by adhesion and also by interaction between the swollen granules.

The high swelling experienced by the granules — up to 30 times or even more — makes the granules so susceptible to a high stirring torque, that abrupt changes in rheological behavior occur and special methods of measurements are needed. A number of instruments have been devised to follow the rheological behavior. Some of the more common ones are described below.

1. Redwood Viscometer

This instrument is based on the measurement of time required for a specific volume of the paste to pass through an orifice. It consists of a heavily silver-plated oil cup mounted onto a bright chrome-plated water bath. The base of the cup which is concave in shape, has a central hole, which can be closed or opened with a ball valve. The level to which the solution is to be filled into the cup is given by an index fixed to the inside wall of the cup. The water bath is provided with a heating coil and a stirring arrangement. The solution whose viscosity is to be measured, is brought to the temperature of the water bath and filled up to the index, keeping the orifice closed with the silver-plated valve. A graduated glass vessel is kept below and the valve is opened and simultaneously, a stop watch started. The time taken for a definite quantity of the solution to be collected in the vessel below is noted.

The instrument is simple and suitable for measurement of viscosity of concentrations of starch. The Indian Standard Institution (ISI) has accepted this instrument for determining the viscosity of starches including cassava. According to ISI specifications, 50 mℓ of a 2% cassava starch solution at 75°C should require a minimum of 44 sec for passing through the orifice.[20] In comparison, maize starch of equal concentration requires only 35 sec, while potato starch requires a much higher time. Modified starches can also be tested by this instrument.

2. Scott Viscometer

A specially constructed water bath is maintained at its boiling point with live steam. The Scott cup is accurately machined and possesses polished orifices and is provided with an overflow to ensure uniform sample size. The flow of a 100 mℓ sample is followed by noting the time in seconds. This machine also gives reproducable results and is fast and simple.[44]

3. Brookfield Viscometer

The Brookfield Synchro-Lectric viscometer consists of a rotating cylinder or disc coupled through a torsion spring to a constant speed-synchronous motor. The spring offers a restoring force to neutralize the resistance created by the sample viscosity to the rotation of the spindle. The torque so produced is read out from a dial. The units are given in centipoise (cP). The instrument can be used to measure apparent viscosities from 1 to 64 mcP. Other properties like dilatancy, thixotropy, and effect of concentration of starch, cooling time, and temperature and shear on the viscosity can also be monitored using the viscometer.[45]

Elder and Schoch[46] have compared various starches using the Brookfield viscograph and found that corn has a higher shear stability compared to potato, which is more stable than cassava starch. It was also found that by cross-linking using phosphate, the shear stability of sorghum starch could be enhanced several times.[47]

4. Brabender Viscoamylograph

This is the most common instrument used, the world over, for characterization of starches. It is a rotational instrument which can continuously monitor the viscosity while cooking and cooling the starch paste. The rotating sample cup is provided with a number of fixed vertical pins and is driven at a constant speed by a synchronous motor. A circular metal disc with several pins projecting downwards into the sample serves as the sensing element. When the viscosity of the sample increases, the force exerted by the sensing element is dynamically balanced by a calibrated tension spring. This causes an angular deflection of the sensing elements shaft and seconded on a strip chart recorder.[48,49]

An electrically heated air bath is provided at the base and heats up the cup, while for cooling, a coil projects into the cup. Heating and cooling are controlled by a mechanically operated thermoregulator which automatically adjusts the heating or cooling to 1.5°C per minute or keeps the temperature constant. It is also possible to obtain absolute viscosities by calibrating with standard oils.

The procedure for determining the viscosity consists of weighing accurately an appropriate amount of starch with adjustment for moisture content and suspending the starch in about 250 mℓ distilled water at 25°C. The mixture is stirred to obtain a lump-free suspension and transferred to a 500 mℓ standard flask and made up to volume with distilled water at 25°C. A pH adjustment may be done, if the pH is below 6.0 or above 8.0. The slurry is transferred quantitatively into the sample cup of the instrument, the sensing element is placed in position, the instrument head is lowered into the operating position, and the thermoregular is adjusted to 25°C. The instrument is started and the temperature is increased by 1.5°C per minute, until it reaches 95°C. At this stage, the temperature is maintained at 95°C for a fixed time while stirring and then cooled to 50°C at 1.5°C per minute, maintained at 50°C for a fixed period, and further cooled. All the temperature controls are carried out automatically by the instrument controls.

A typical Brabender curve for a starch paste gives five points of interest. They are:

1. Peak viscosity represents the highest viscosity the starch paste can reach.
2. Viscosity at 95°C, which in relation to the peak viscosity, gives an idea of the ease of cooking of starch.
3. Viscosity after cooking at 95°C for a certain period, reflects the stability or breakdown of the paste.
4. Viscosity of the paste after cooling to 50°C is a measure of the setback produced by cooking.
5. Final viscosity after stirring for definite period at 50°C indicates the stability of the cooked paste.

When the Brabender viscosity curves of various starches are plotted, wide variations are observed between them. Wheat, corn, and rice starches show an increase in viscosity much later than cassava and potato. These starches also exhibit lower peak viscosity compared to cassava or potato starches. However, the breakdown observed in the case of cassava and potato starches is much larger as indicated by lower viscosity at 95°C compared to peak viscosity. As the temperature is maintained at 95°C for 30 min, cassava starch shows a slow fall and reaches almost half the peak viscosity value, while the cereal starches show only nominal breakdown. The setback observed for cassava starch is much lower than that for the cereal starches, so that on cooling, the final viscosity is higher for cereal starches. However, this is not the real viscosity, as it has undergone retrogradation.

The early gelatinization, higher peak viscosity, and higher breakdown of cassava starch indicate that the associative forces between the starch molecules in the granules are relatively weak compared to cornstarch. The water molecules are able to penetrate the starch granules much easier, and the granules swell enormously, leading to weakening of associated forces, which in turn makes them susceptible to breakdown. This breakdown is responsible for the long, cohesive nature of the cassava starch paste.

As is evident, strengthening of the granules can help in reducing the breakdown in viscosity. A comparison of cross bonded and native cassava starch indicates that the former shows lower peak viscosity and lower viscosity breakdown. However, if the cross-bonding exceeds an optimum level, the granules may not gelatinize completely or at all.

Various physical and chemical treatments of starch affect the viscosity. For example, heat moisture treatment increases the strength of the associate forces leading to higher pasting temperatures and lower viscosities. Steam pressure treatment has been found to reduce the peak viscosity considerably, depending on the time of treatment and steam pressure used. The peak viscosity of 6% cassava starch paste falls from an original value of 600 BU to 30 BU on steam pressure treatment for 150 min at 20 psi. Along with the reduction in peak viscosity, the viscosity breakdown also shows a fall, indicating that the swelling is reduced by increased associative forces brought about by bombarding with steam. In the case of potato starch, the heat moisture treatment has been found to convert the X-ray diffraction pattern from B to A, showing that the treatment compresses the starch molecules to a more compact structure. Cassava starch being of an A pattern does not undergo any change.

The addition of chemicals also changes the viscosity characteristics. The effect of chemicals on starch gelatinization is important because during production of various products, many ingredients are added. A knowledge of the effect of the ingredients will be very valuable to obtain the desired effects. For example, the effect of sugars is important in the preparation of pie fillings, creams, puddings, and sweet sauces. Sucrose has been known to impede the swelling of starch. Small additions have been found to increase peak viscosity. Above 20%, the peak viscosity is reduced.

The pH of most foods lies between 4.0 to 7.0, and not much effect on viscosity has been observed. In the case of foods with much lower pH, like salad dressings and lemon and fruit pie fillings, cross-bonded starches may have to be used. Citric acid has been found to increase the maximum viscosity, but the paste breaks down faster.

The effect of salts on viscosity has also been examined by various workers. Sodium chloride in concentration as low as 10^{-5} N has been found to reduce the viscosity of potato starch. The viscosity was reduced tremendously, when the strength of sodium chloride was increased to 0.001 N. In contrast, cornstarch and wheat starch showed only a very minor effect. Our experiments with cassava starch also showed only a slight decrease in viscosity. The noticeable effect on potato starch has been attributed to the presence of polyelectrolyte phosphate groups. In contrast, the sodium form of potato starch, obtained by ion exchange treatment with sodium chloride, produced a higher peak viscosity and the calcium starch exhibited a lower peak viscosity.

Proteins have also been found to influence the peak viscosity of starches, but the effect varies with the type of starch used, the nature of the protein, and the presence of other ingredients.

Fats and surfactants also affect the viscosity depending on factors such as:

1. The type of concentration of amylose
2. The ratio of amylose to fatty acid
3. The location of fatty acid in the homologous series

It was also found that though fatty acids caused an increase in viscosities, the increase in the case of tuber starches was lower than that obtained for cereal starches. Krog[39] has studied the effect of various types of surfactants on cassava starch viscosity and found that calcium steroyl lactylate and diacetyl tartaric acid ester of monoglycerides slightly reduced the peak viscosity of cassava starch without stabilizing the viscosity, while GMS at 0.5%

concentration stabilized the viscosity (Figure 3). The effect of different types of surfactants on cassava starch properties has been examined by Moorthy[40] and it was found that the viscosity is reduced slightly by the anionic surfactants, sodium palmitate and stearate, while sodium lauryl sulfate and cetyl trimethylammonium bromide increased the peak viscosity. Whereas sodium palmitate and sodium stearate stabilized the viscosity, the other two surfactants imparted only low stability to the starch viscosity. The results indicate that the complexes with the latter surfactants are quite unstable. It has also been demonstrated by various workers that the size of the hydrophilic and hydophobic regions of the surfactants play an important role in determining viscosity stability.[39,40]

K. Swelling Power and Solubility

These are important properties which vary very widely between different types of starches. The swelling power may be defined as the maximum increase in volume and weight which the starch undergoes when allowed to swell freely in water. The determination of swelling power is done by suspending a weighed quantity of starch in distilled water in a centrifuge bottle and heating it for 30 min at the specified temperature, while stirring to ensure a homogeneous suspension of the starch. The stirring should be gentle, otherwise the granules may disrupt, leading to lower value for swelling power. The sample is then centrifuged, the aqueous supernatant removed, and the weight of the swollen sediment determined. The swelling power is calculated as the weight of sediment per gram of starch used, with a correction applied for the solubles. The swelling volume is the volume of the undissolved sediment obtained on centrifugation. Solubility is the weight of the starch solubles and is determined directly by drying and weighing an aliquot of the supernatant. Swelling and solubility patterns are usually obtained by plotting the values found at 5°C intervals over the pasting range of the starch. Leach et al. have found that each species of starch showed characteristic swelling and solubility patterns.[36] There is a direct interrelationship between swelling and solubility, as indicated by the similar curves obtained for each type of starch. On plotting swelling power against percentage solubles, an almost straight line was obtained for different starches confirming the close relationship between the two properties. This is understandable when the basic factors governing swelling are examined. The swelling behavior of starch is dependent on the strength and nature of the micellar network within the starch granules, which is correspondingly dependent on the nature and strength of associate forces within the granules. The various factors which determine the associative forces include (1) ratio of amylose to amylopection, (2) molecular weight of the fractions, (3) molecular weight distribution, (4) degree of branching, (5) conformation, and (6) length of the outer branches of amylopectin molecules that can partake in associative linkages.[36] The presence of naturally occurring noncarbohydrate impurities also affects the associative forces.

The normal starches fall into 3 groups, which follow the order of degree of association. Thus, the cereal starches with the highest degree of association have the lowest swelling power and solubility, followed by root starches and then tuber starches. Many cereal starches also show two-stage swelling and solubility patterns, reflecting the presence of two types of internal associative forces, which probably exist in the crystalline and amorphous regions of the granules.[3] A comparison of the waxy and nonwaxy varieties of rice, sorghum, and corn starches shows that the higher amylose content leads to lower swelling, probably because the linear molecules reinforce the internal network. Cassava starch shows a higher swelling power, single stage swelling, and a higher solubility, again reflecting the lower strength of associative forces in the granules. Potato starch shows enormous swelling due to weak internal bonding. In addition, the presence of ionizable esterified phosphate groups assists swelling by means of mutual electrical repulsion.

The presence of foreign ingredients in starch also affects swelling. When corn starch was defatted, the swelling was much faster and the pattern became one stage. The natural fatty

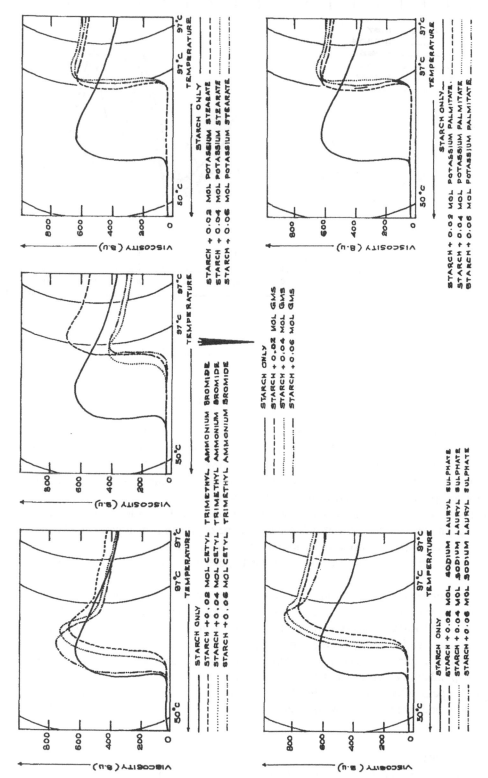

FIGURE 3. Effect of surfactants on cassava starch viscosity. (From S. N. Moorthy, *J. Agric. Food Chem.*, 33, 1227, 1985. ©1985, The American Chemical Society. With permission.)

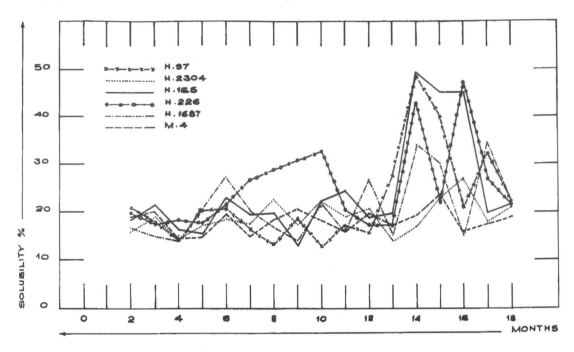

FIGURE 4.   Effect of age and variety on solubility of cassava starch.

acids present in corn starch probably form an insouble complex with the linear fraction leading to restricted swelling. Addition of surfactants also affects the swelling power of starches. Potassium palmitate and stearate reduced the swelling power of cassava starch, whereas sodium lauryl sulfate and cetyl trimethyl ammonium bromide increased the swelling power.[40]

Chemical modification also affects the swelling power and solubility of starches. Acid modification or hypochlorite oxidation lead to hydrolytic or oxidative cleavage in the more accessible intermicellar regions, thereby weakening the granule network. This results in fragmentation of granules and higher swelling and solubility. In contrast, starches of increased viscosity are produced by esterification or etherification. In these reactions, the internal bonding is weakened by introduction of substituent groups, which leads to increased swelling. The change in these properties is also dependent on the degree of substitution and the nature of the substituent group introduced, particularly if the group introduced is an ionic one. Cross-linking increases the bonding between starch molecules leading to lower swelling and solubility.

The swelling power and solubility of cassava starch are also dependent on varietial differences, environmental factors, and the age of the crop as evidenced by detailed studies by Moorthy and Ramanujam (Figure 3). It was found that swelling volume and swelling power of a few varieties were depressed tremendously when the maturity period was exceeded, and a simultaneous increase in solubility was noted. The low swelling accompanied by the high solubility obtained indicates the very weak associate forces in these varieties, and this happens after 8 to 9 months. It was found that there is no correlation between granule size, and molecular weight of starch with the swelling power and solubility (Figure 4).

L. Paste Clarity

A suspension of starch in water is opaque, but gelatinization of the starch granules increases the transparency of the solution. A quantitative relationship between the swelling of starch and paste transmittancy has been established by Kite et al.[50] They found a linear correlation when light transmission was plotted against swelling power and it was obtained for all unmodified starches, including cassava. Thus light transmittance can be considered as a direct measure of granule swelling. Visual estimation of paste clarity, however, is related to light reflectance of the paste, which reflects the differences in optical homogeneity within the granules with respect to uniformity of hydration.[50] Paste clarity is related to the state of dispersion and retrogradation and also increased paste clarity. It therefore follows that cassava starch with high swelling power and also lower retrogradation tendency should have a higher paste clarity, and this is experimentally found to be true. Hence, cassava starch is preferred in foods which require good clarity, e.g., in pie fillings and gravies. The presence of other ingredients in food also affects the paste clarity, e.g., sugars increase the paste clarity, GMS reduces the clarity, while sodium lauryl sulfate and cetyl trimethyl ammonium bromide increase the paste clarity.[39,40]

M. Sol Stability

On cooling a starch paste, the starch molecules become less soluble and tend to aggregate and partially crystallize. With dilute pastes, the gel precipitates out, but at higher concentrations, usually encountered in foods, the polysaccharide molecules form three-dimensional gel networks. This network encompasses both amylose and amylopectin molecules, but the alignment and crystallization of amylopectin are partially inhibited by its ramified structure and are restricted to the outer branches. These crystalline areas, found within the swollen granules, decide the strength and rigidity of the gel. Sterling[51] has explained that the elasticity of the starch gels is due to the ability of the gel to be stretched to a certain extent without breaking, and has attributed this property to the portion of macromolecules that lie in the amorphous regions between the crystalline micelles. This separation becomes even more noticeable when the gel is frozen and thawed. Although gel formation is desirable in many food products, the changes associated with retrogradation like "curdling" of frozen sauces upon thawing, or staling of baked goods, are not desirable. Similarly, the formation of a skin on the surface of starch pastes is due to retrogradation. Whereas retrograded amylopectin can be brought back to a disposed state by heating to 50 to 80°C, amylose, once retrograded, can never be dispersed, even by autoclaving.[53]

The gel stability of cassava starch is much higher compared to cereal starches and hence is preferred for many food products especially when the food has to be stored for a long time. In addition, the higher paste clarity and bland flavor are desirable properties of cassava starch.[52,54]

For some foods, storage at low temperatures is required. For example, a fruit pie is stored nearly frozen, but it will be frozen and unfrozen many times as required for consumption. It is clear that during these operations, starch paste must not breakdown, i.e. the sol stability has to be maximum. Though cassava starch has a high degree of sol stability, the outer branches of the starch amylopectin units can still form some degree of parallel association leading to gel separation, especially when subjected to repeated freeze-thaw operations. This may be prevented by making the branched structure of amylopectin starch more branched by introducing side chains. The derivatization is also advantageous in that it also increases paste clarity and lowers the gelatinization temperatures. Knight has shown that cross-linked and acetylated cassava starch remained a soft paste after six cycles of freeze-thaw operations.[54] Moorthy has also found that the sol stability of cassava starch can be considerably enhanced by derivatization.[37,38] The addition of amylose complexing agents also increases the sol stability.[38,39] Various surfactants increased the stability differently and sodium lauryl sulfate and cetyl trimethyl ammonium bromide were found to be the best.[38]

VII. APPLICATIONS OF CASSAVA STARCH

A. Starch in Textile Industry

There are three general areas in which starch is mainly used by the textile industries.

1. Textile Sizing Operations

This involves coating the warp yarn with a smooth film of starch in order to enable the yarn to withstand the abrasive and flexural stresses which the yarn is subjected to during the weaving process.

An enormous amount of starch is used in textile sizing operations. Yarns spun from staple fibers such as cotton are slashed by 10 to 30% with starch solutions to improve the strength of the yarn and the increased stiffness facilitates handling in starting the warp through the loom. The size films are applied as a thin coating on the surface of the spun yarn and should be applied so as to penetrate the yarn only far enough to provide satisfactory adhesion, otherwise it will be too stiff. Normally, the amount of size added to spun yarns is 5 to 15% of the weight of the yarn, but it depends on the viscosity and type of the starch used. The yarn filaments are cemented to prevent the formation of "fuzz balls" which occur when a single filament breaks and is pushed back along the body of the yarn and the broken filaments accumulate to cause sufficient entangling to stop the loom. Here the size must penetrate the yarn completely and the amount of size used is 2 to 5% of the weight of the yarn.[55]

Specific properties required for providing best sizes are as follows:

a. Low Cost

The material used for sizing is removed after the weaving operation and since a large quantity of material is used to bring about a temporary change in yarn properties and it cannot be recovered, it has to be inexpensive.

b. Easy Year-Round Availability

The starch should be available in good quantity throughout the year.

c. Easy Preparation

Size solutions are prepared without much mechnical expertise by unskilled workers who receive only minimum supervision.

d. Uniform Viscosity and Solids Content

The amount of size on the slashed yarn should be uniform, in order to ensure uniform weaving performance, and to reduce the fluctuations in the weight of the greige fabrics.[55] Viscosity and solids content also affect the penetration of the size into the yarn and hence also the adhesion of the size. Therefore, these properties have to be maintained as uniformly as possible. The temperature of the size solutions must be kept constant, since viscosity is increased by reducing the temperature.

The quality of starch obtained from different batches has to be maintained uniformly so that the viscosity and solids content remain more or less the same between the various size solutions.

e. Nonsticking Nature

The sized yarns are dried with hot air or in contact with rotating steam heated cylinders. The sized solutions should not be tacky, otherwise the yarns will stick to cylinders or the adjacent yarns will tend to adhere to each other.[55]

f. Low Foaming

Since sizing operations are carried out at maximum speeds, chances of foaming exist, which can cause nonuniform size deposition. A pH of 5.5 to 6.5 (slightly acidic) is given to starch sold to textile industries. Acetic acid or zinc chloride may also be added to the size during cooking.[56]

g. Resistance to Microbiological Damage

Size solutions may be held overnight or stored for weekends, and hence have to be resistant to microbial action. Mildicides or fungicides may be added.

h. Low Pollution Problems

The large amount of starch found in the waste from sizing operations causes a large biological oxygen demand (BOD) in the effluent and hence other chemicals like carboxymethyl cellulose may be added to reduce pollution problems.[57]

A size film obtained from the size solution should have the following properties, viz. — adhesion to fibers which occurs probably by formation of hydrogen bonds between starch and cellulose; hardness to resist abrasion, but softness to be stretched; ease of removal during the wet processing after sizing; resistance to heat damage, etc.

Cassava starch has some desirable properties along with some undesirable properties. The lower price of cassava starch is the main attraction; generally, cassava starch is available cheaper. The starch film obtained is more transparent and this is important for colored goods where the color appears bright. The film obtained is also more flexible compared to cornstarch film.[58] Another advantage lies in its better keeping properties in view of its lower tendency to gel. The starch also gelatinizes much easier and faster compared to cereal starches. Cassava starch completely gelatinizes on boiling for 2 hr but cornstarch, even with thorough boiling, does not result in complete gelatinization.

A number of chemicals are also added to sizes to improve the properties. These include film modifiers like fats, surfactants, and glycerin, and lubricants like paraffin wax and synthetic polymers. The machinery and processes used for making the size solution have been described by Compton and Martin.[55]

The type of starch used varies in different countries depending on properties and availability. Sago starch is more popular in the U.K., corn starch in the U.S.A., while in India, both cassava and cornstarches are used.

However, cassava starch also possesses some undesirable characteristics. Though it possesses a high viscosity, the value breaks down very rapidly upon boiling for a period of time. This leads to an uneven size on the yarn. The yarn produced at different times will have varying strengths leading to problems in weaving. In addition, the high breakdown of granules leads to a cohesive texture of the paste, which leads to a tendency to stick to the machinery and drying cylinders. A 5% paste of cornstarch exhibits a peak viscosity of 200 to 240 Brabender Units (BU) under standard conditions and this value is maintained for some time, until retrogradation starts.[36] However, cassava starch, at the same concentration, shows a peak viscosity of over 600 BU, but the viscosity falls rapidly to about half this value in 15 min during stirring and heating. The swollen cassava starch molecules in the granules, being linked through weaker associate bonds, fragment into smaller units and this leads to the long cohesive nature of the paste.

Thus cornstarch has an advantage over cassava starch in the case of viscosity. The problem of unstable viscosity can be overcome by various methods. These include blending with other starches, physical treatments, addition of chemicals, and chemical modification of the starch. For example, a mixture containing one third cassava and two thirds cornstarches gave good results with 85 and 125-count yarn denims.[58] Moorthy[59] has found that various cassava-maize blends show a reasonable peak viscosity, good viscosity stability, clarity, and sol stability. Steam pressure treatment has been found to be effective in lowering peak viscosity and correspondingly reducing the viscosity breakdown.[60] A 6% paste of cassava starch treated for 90 min at 15 psi pressure exhibits a peak viscosity of 100 BU and a breakdown of 20 BU. The increased resistance to swelling and hence breakdown may be attributed to the strengthening of associative forces, which might have been brought about by the steam pressure. The addition of surfactants also brings about the stability of viscosity, probably by forming stable complexes with starch molecules preventing them from breakdown.[39,40] Cross-linked cassava starch has been found to have extremely good viscosity stability.[61,62] Srivatsava and Patel[62] have standardized the conditions for production of cross-linked cassava starches using phosphorus oxychloride, epichlorohydrin, and sodium trimetaphosphate for use in the textile industries. If these modified starches can be manufactured cheaply, they will be in great demand. Though oxidized starch is not so widely used in warp sizing, its high fluidity, stable viscosity, and flow properties at a high solid level allow for greater add-on to yarn and provide abrasion resistance.[63] The starch acetates are mainly used in warp sizing, because of good yarn adhesion, tensile strength, and flexibility, besides being easily removed due to the solubility of its film.[64] It is also blended with poly-(vinyl alcohol) in warp sizing to lower the cost.[63]

Low substituted hydroxyethyl starches are used — alone or blended with PVA — in warp sizing.[63] In addition to cotton yarn, polyester-rayon and polyester-cotton yarns can be sized using a mixture of 50 to 90 parts of starch monophosphate and 10 to 50 parts of PVA or polyacrylate.[65] Cationic starches and amphoteric starches are used as warp sizing agents because they provide lubrication and abrasion resistance.[66]

2. Textile Finishing Operations

Most fabrics are subjected to finishing operations to modify their appearance, to change the stiffness or hand of the fabric, and to add weight.[67] The finishing may range from a treatment with very weak solutions of starch, to a "back-filling" finishing method for which a heavy viscous "starch mix" is used. Since the finish using starch is temporary, it is used only with inexpensive fabrics. Because the fabrics finished with starch alone are stiff and brittle, usually film modifying material are also added. To obtain a stiffened finish, starch may be used in conjunction with the thermoplastic or thermosetting resin and it gives a permanent change to the fabric properties.[55] The fabric in the last stage of finishing is immersed in a dilute solution of cooked starch, squeezed to remove the excess solution and dried on steam-heated cylinders, or the fabric may be passed through a calender. In back-filling, the starch is applied along with a filler to the back of the fabric. The stiffness and the opacity of the fabric are increased, since the interstices of the weaves are filled by the starch. Various types of equipments are used for this purpose.

Radley has compared the various starches for this purpose and cassava starch size has been found to be softer and more transparent than cornstarch, while finishings from the starch are more flexible and tougher.[68] In view of this advantage, cassava starch is often preferred to cornstarch for finishing purposes. Oxidized starches may be used in back-filling, since the lower viscosity helps it to penetrate the fabrics to a much higher extent.[63] Starch acetates are used, in combination with thermosetting resin, as inexpensive finishes, providing weight and hand to the fabrics. They are also used in finishes for interliners to give stiffness.[63]

Hypochlorite oxidized starches may be used for finishing printed fabrics, particularly cotton fabrics, as the less opaque film will not dull the colors.

3. Textile Printing Operations

For printing, the usual procedure is to add dyestuffs or their solutions to a paste prepared from starch, gum, albumin, etc. These pastes are then applied to the cloth by different methods. Starch, which functions as a thickener, is used to give the printing paste the consistency necessary to produce a clean, sharply-defined pattern of color.[55] Since starch pastes sometimes react with the dyes used and are affected by acids and alkalies, gum tragacanth is also added to the paste. The use of starch derivatives is also helpful in preventing any reaction with dyes or other chemicals.

The starch solutions for this purpose are prepared similar to sizes, but the viscosity is higher, since the printing is carried out at room temperature. The dye and other chemicals are added to the cooled, cooked solution of starch.

Among the different starches used for the purpose, wheat starch is considered most suitable, but cassava-corn or cassava-wheat blends give nice, smooth working properties.[68] A typical formula consists of 630 parts cornstarch, 320 parts cassava starch, 4000 parts water, 2000 parts of 8% gum tragacanth mucilage, and 480 parts acetic acid.[68]

Starch is also used along with glue and wax in the solution applied to sewing thread in the final finishing operation. The thread is passed over a roll partially immersed in a trough containing the hot solution of starch. The thread is then passed over rotating brushes and dried in a stream of hot air. The finish thus formed improves the luster of the thread and reduces friction during the sewing operation. One of the major drawbacks which many industrialists reportedly face in using cassava starch for textile sizing and finishing purposes is the high inconsistency in quality between different samples. The viscosity values vary widely and many batches show Redwood viscosity values well below the 44 sec required by ISI standards.[20] This makes the sizing nonuniform, leading to uneven weaving. The nonuniformity in quality can be attributed to a number of factors such as improper processing, difference in varieties, different types of machinery used for extraction, and the nature and quality of the water used for the separation of the starch. However, the most important factor affecting the quality is improper processing. It has been generally observed that improper drying leads to microbial damage which causes a lowering of viscosity and color. The starch obtained from dried chips also shows lower quality. In order to ensure that quality is maintained, it is necessary to modernize the small scale starch manufacturing units, use high yielding varieties, harvest at correct maturity, and use modern methods of storage and transportation.

B. Starch in the Paper Industry

Though the basic component of paper production is cellulose fiber, a number of other materials are used to process and modify the finished products to suit various end uses. It is estimated that for the processing of 50 million tons of wood pulp, approximately 12 million tons of chemicals and pigments are used.[69] The Technical Association of the Pulp and Paper Industry (TAPPI) reported that the 1976 estimate of natural binders used by the U.S. paper industry exceeded 1.2 billion kg, of which starch accounted for 97%.[70] Jones[71] has found that 60% of the unmodified starch used in the U.S. is in the paper industry, while approximately 50% of modified starch is used for the same purpose.

However, in India, the amount of starch used by paper industries is very small (less than 10%), but it is forecast that the demand for starch in the paper industry will rise.[72]

Starch is used in paper and paper-board manufacture in four major operations. The function of starch in different stages and the desirable characters are discussed below.

1. Wet End Application of Starch

The purpose for which starch is used is to increase the tearing and bursting strength of finished paper, improve the retention of fibers in the wire mesh conveyor in the continuous manufacturing process, and to allow higher retention of fillers in the final product and reduce BOD of the effluent.

The cellulosic fiber requires strengthening before being made into paper sheets. One process carried out is "refining". However, in addition, it is necessary to add binders to improve the properties of the finished paper and starch is mainly used for this purpose. The process of the addition of starch to the wet end of the paper machine is called beater sizing. The starch acts to increase paper strength, to lay surface fuzz, to increase stiffness and rattle, and to allow inclusion of inorganic fillers.

Various starches including cassava starch have been compared in their ability to bind cellulose molecules.[73] An ideal starch is one which gives maximum starch retention at minimum cost and has low deleterious side effects such as pigment fallout. Though no particular starch meets all the requirements, potato, cassava, sorghum, and cornstarches are mostly used.[74] For loosely constructed sheets, poorly dispersed pastes containing large, swollen, unbroken granules are mostly used. The efficacy of the large granule root starches as wet-end adhesives for low-density sheets has been explained on the basis of the "spot welding" theory of fiber-starch-fiber bonding. According to this theory, the fiber interstices in low density sheets are relatively large, and large swollen granules having a size of approximately 20 μm, fit in correctly to produce a strong bond.

The amount of starch added and the amount that is retained also vary between the different types of starches. Waters has found 26.7% retention for potato and 9.45% for corn.[75] It can be expected that for cassava starch the value should be between these two values. The low retention of starch can be attributed to the anionic character of both cellulose and starch. Hence cationic starches are more efficient, being retained much more. The cationic starch has a high electrochemical affinity for the negatively charged cellulose fibers and this results in almost complete adsorption of the starch. Nissen found that 1% cationic starch imparts an efficiency equivalent to 3% unmodified starch.[74] It behaves as an ionic bridge between cellulose fibers and mineral fillers and pigments. Cationic starches are preferentially adsorbed on the pulp fines and this leads to increased retention of fines and these fines along with the long fibers form a cohesive network which results in increased strength. However, alum, which is used as an additive, affects the efficiency of cationic starch, and hence improved cationic starches were developed, .e.g., starch quarternary ammonium ether with a DS = 0.033, and an anionic poly(acrylamide) starch derivative improve pigment retention even in the presence of alum at pH 4 to 5.

Workers at ATIRA, India have shown that cationic cassava starch of good quality gives reaction efficiences of around 95%, and the products exhibit better performance as wet-end additives in the manufacture of kraft paper when compared to products made from cornstarch.[72]

Oxidized starch, often introduced into the repulped, coated paper (broke) has been reported to have negative effect on pigment retention by acting as a dispersing agent.[76] But starch acetate does not have this drawback. Hydroxyethyl starch can be conveniently added to the wet end of the paper making machine, since it has a lower gelatinization value and can gelatinize when the paper sheet passes over the drying rolls. The swelling helps in increasing the internal strength.[77] Amphoteric starches produced by phosphorylation of cationic starch are effective as wet-end additives.[78]

Dialdehyde starches also have tremendous potential in wet-end application for better retention at lower cost and for obtaining a softer and more water absorptive tissue. The wet strength obtained by using diadehyde starch is less permanent compared to that obtained by using resins, and this helps in better broke recovery and faster biological degradation of the paper tissues in septic tanks and similar sewage systems.[79] The process for producing wet-strength paper using dialdehyde starch consists of the addition of a finely dispersed dialdehyde starch to cellulose pulp. An aqueous dispersion of 3% dialdehyde starch is prepared in water containing 0.45% sodium metabisulfate by cooking for 30 to 40 min at 95°C. The dialdehyde starch is broken down to a molecular weight range of 0.3 to 5 million and produces a negatively charged hydrocolloid because of the presence of sulfonate radicals in the structure. Hence it will not be well-attached to the negatively charged cellulose. Therefore, the pulp is treated with a small quantity of cationic starch and to this aqueous dispersion, the hydrocolloid is added at a 0.25 to 0.75% concentration at pH 4.0.

To improve the dispersibility and retention, cationic properties may be built into the dialdehyde starches by reacting with betaine hydrazide hydrochloride. The improvement in wet tensile strength and dry tensile strength by addition of a small quantity of cationic dialdehyde starch is quite good. The mechanism of wet strength increase by dialdehyde starch is presumably accomplished by cross-linking of cellulose through acetal formation between carbonyl groups of dialdehyde starch and the free hydroxyl groups of cellulose.[79]

Mehltretter has listed the advantage of dialdelyde starch as a wet-end agent for paper:[79]

1. High and temporary wet strength at low levels
2. Almost maximum wet strength obtained without after-cure
3. Ease of broke recovery
4. Biodegradability of the paper
5. Stability of treated papers
6. Ease of preparation of dispersions
7. Production of improved wet and dry strength paper from a variety of pulps

Considerably large quantities of wet strength resins are presently used and part of this can be conveniently replaced by dialdehyde starch.

2. Size Press Applications of Starch

Size press application is known as tub sizing or surface sizing and was originally carried out by immersing the sheet in a tub of starch solution, passing the starch-saturated sheet through press rolls and drying the web. The modern starch mills use size presses which are of vertical or horizontal design. Starch sizing helps to improve the appearance and the erasability, to inhibit ink penetration, to form a hard, firm surface for writing or printing, to reduce surface fiber picking, and to prepare the sheet for subsequent coating.

The type of starch used depends on availability, and all starches such as cassava, potato, and cornstarches are used in their native forms for the purpose. Viscosity stability of oxidized starch dispersions make oxidized starch quite suitable for the purpose, but recently, the introduction of on site enzyme or thermal-chemical converted starches has reduced the use of oxidized starch for paper sizing purposes.[63] Starch acetate finds wide use in surface sizing because of improved printability and other properties, e.g., providing low and uniform porosity, surface strength, abrasion resistance, oil holdout, solvent resistance, and capacity to adhere stray fibers to the substrate.[82,38]

Hydroxyethyl starches of low molar substitution are useful in surface sizing of paper — providing strength, stiffness, and ink hold out.[84] Starch phosphates can be employed in emulsions with ketene dimers as internal sizing agents for paper, and the FDA has approved a combination of phosphoric acid and urea for starch modification in surface sizing of paper.[85] Cationic starches are finding increased importance in emulsifying synthetic sizing agents of paper. The cationic starches, because of their irreversible ionic attraction to cellulose fibers, penetrate less and also reduce BOD in the effluents.[83,86,87] A further modification of the cationic starch like oxidation, acid or thermal catalyzed conversion, can improve the sizing efficiency. The film-forming property of high amylose starches has been combined with cationic modification to yield an improved surface sizing starch.[88] ATIRA has found that cationic cassava starch proved to be an excellent surface sizing agent for paper,[72] but the quality of the starch used has to be uniform to ensure the best results.

3. Colender Application

In this process, the paper passes around a series of steel rolls, one or more of which may transfer a starch film to the paper surface. The type of starch used depends on the fiber furnish, machine conditions, and end-use requirements. For heavyweight papers and liner boards, native or thin boiling starches are used in concentrations of 2 to 5%.[74] In cylinder boards, a low viscosity starch with good film-forming ability is mostly used, usually in the form of hydroxyethyl starch or oxidized starch. In base coat surface preparation, medium or low-viscosity oxidized or hydroxyethyl starches are used. Similarly, for control of curl in paper, low viscosity, oxidized starches are utilized to film the lower surface. The major conditions for correct application are that the starch is thoroughly cooked and that concentration and temperature are correctly maintained.

4. Paper Coating

Starch functions are used as an adhesive in pigmented coatings for paper and paperboard. Starch is used in view of its easy preparation, high adhesive strength, cheapness, stable viscosity, and stability to extended treatment.

The most important characters which determine the property of starch-based coating colors are the water holding capacity and viscosity, which can be manipulated by the type of starch used.

Mostly hypochlorite-oxidized starch is used in view of its high fluidity, good binding, and adhesive properties which make it effective in high solid-pigmented coating colors. The factors to be taken into consideration for its use are its compatibility with the pigment, and that it should not adversely affect the water-holding capacity and rheology of the coating color.[89] Low substituted hydroxyethyl starches can also be made use of as a binder in pigmented paper coatings, because they provide good leveling and viscosity stability in the coating color.[90] The high water holding capacity controls binder penetration, and provides a high binder strength for the pigment and adhesion to the base stock. The resulting coating provides good printing quality. These starches can be used for the preparation of wet-rub resistant coatings with glyoxal, and can be admixed with synthetic polymers in clay coating.[91-92]

Starch phosphates have good dispersant properties for clay-satin white coating colors.[93] Combined with white satin pigment, urea-phosphate starches can be used in paper coating. Rutenberg and Solarek[63] have described the preparation of low viscosity urea-phosphate cassava starch. A mixture of 1000 parts starch, 75 parts orthophosphoric acid and 140 parts urea are heated at 126° for 45 min to obtain a product with a viscosity of 34 cP at 25°C at 15% concentration.

It has been found that an excellent wet-rub resistance of 90 to 91 in Adams' test is obtained for coating colors consisting of the above starch, satin white, and china clay.

Cationic starch provides increased strength as a coating binder because of the electrochemical binding of clay to the fiber. A starch product containing amine and carboxyl groups is claimed to yield good binding and shock-free clay dispersion.[95,96]

Thus, starch finds use in modified and unmodified forms in various stages of paper production and more and more types of starch are being used to suit different needs. The potential for cassava starch in this field is good.

C. Starch in the Adhesive Industry

Starch was used for the preparation of adhesives 6000 years ago by the ancient Egyptians. Starch adhesives became more common during the Industrial Revolution and the advent of postage stamps and gummed envelopes, followed by photography and safety matches, increased the demand for adhesives. In the packaging industry, the animal glues which formed the most common industrial adhesive until 50 years ago have been replaced by vegetable adhesives.

Because of the various uses of starch adhesives, the most important prerequisites are good flow characteristics and low setback so that they can be pumped through narrow pipes and can be applied by transfer rollers:

The availability and properties of various starches decide their use in different areas. In Australia, wheat starch dominates, while it is potato starch in Europe. Cornstarch is mostly used in North America, while cassava starch is the principal starch in Latin America.[97] A study of the properties of different starches reveal that root and waxy starches yield adhesives with good flow characteristics, while those from cereal starches have poor mobility and are more suitable for purposes where short parts are required, e.g., in the manufacture of corrugated boards. Modified by oxidation or to a thin boiling starch, cornstarch can be also used for other purposes. Wheat starch is usually used as a thick paste in the adhesive base for bill posting and paper bag making. Potato and cassava starches are the most preferred ones, and cassava starch has certain advantages over potato starch. The adhesives are more viscous and smoother working, and fluid, stable glues of neutral pH can be easily prepared and can be combined with many synthetic resin emulsions. Joints made from cassava starch adhesives are considered to exhibit a higher tensile strength than those from potato starch. Cassava paste is neutral in taste and odor, certain potato pastes have bitter properties, while cereal starches exhibit a cereal flavor.[98] However, it is necessary that the starch used for adhesive preparation should be of good, uniform quality.

Generally, adhesives may be classified as glues or pastes. Glues can be defined as those which possess good mobility and stability and can be pumped through pipelines. Pastes have poor or no mobility. Adhesives can be presented either as ready-for-use liquids or as dry powders, and each form has its own advantages and disadvantages. Liquid adhesives have the merit that they can be directly used by the consumer. In addition, a wide range of adhesives can be made available, since the manufacturer can modify the properties by the simple addition of various chemicals. However, these have a short shelf life, are more prone to microbiological damage, and are more difficult to transport.[97]

Dry adhesives can be prepared as needed and have unlimited shelf life, if packed properly. The adhesive dispersion, once prepared, is generally stable for up to a week. The dry adhesives can be classified as cold water or hot water soluble. The former requires less than half an hour for dissolution. The latter is very cheap, but requires some cooking before being dispersed. It may also have to be cooked before applying. There is a tendency to setback, and hence, cooked pastes cannot be easily pumped to different regions. These also possess good storage life. The hot water soluble starches are either a mixture of starch and white dextrin or hypochlorite oxidized starch, or acid-treated thin boiling starches.

The process of production of starch adhesives can be classified according to what type of treatment the starch receives. The simplest starch pastes are produced by cooking the starch in water with the incorporation of only preservatives. These pastes are used in bill posting, bag making, and in tobacco products. The addition of simple salts also improves the properties. A number of workers have studied the effect of various salts at different concentrations and at various temperatures on swelling and gelatinization of starch. Courtonne[98] found that chlorides exerted the maximum effect on gelatinization point. This effect has been made use of by Moller Holthamp who made an adhesive by treating a thick paste of the starch with calcium chloride, and reboiling.[99] Salts act as swelling agents, stabilizing agents, or for imparting transparency, adhesiveness, density, or for increasing viscosity. A few examples of use of salts in the manufacture of starch-based adhesives are given below: To 100 kg starch suspended in 180 kg water at room temperature, 115 kg calcium chloride is added with constant stirring and stirring further for two hr yields a transparent syrupy paste. The addition of 1 kg borax at the end of the process can increase the viscosity. This adhesive which is prepared easily has a higher water content and is used in the manufacture of wallpapers, as a binder for surfacing pigments, metallic powders, etc. Magnesium chloride may be used instead of calcium chloride. Dulac found that the amount of calcium chloride may be reduced by the use of alum.[100]

Borax is also widely used in starch-based adhesives. When borax is added to starch paste, the mass becomes rubbery and nonspreadable, but a workable paste can be obtained by treatment with an acid. Borax-treated starch is used in the preparation of laundry stiffening mixtures, since it allows a better glaze on ironing or colendering.

Radley has described in detail various treatments which can be carried out on starch to obtain adhesives of different properties. These include treatment with alkali, alkaline salts, acids, oxidizing agents, swelling agents, and the addition of chemicals.

As already mentioned, a number of other compounds are added to improve the working properties. The modifying agent used varies according to the properties needed. For controlling consistency, borax, soda ash, or methyl cellosolv are used; for increasing water resistance, incorporation of water soluble components or water soluble precondensation of urea-formaldehyde and resins is used. Formaldehyde or formaldehyde generators may also be used to achieve the same effect. Wetting agents, soaps, resins, protein glues, and rubber latex are added to impart greater penetration power, maximum spreadability, and greater tack.[98] Plasticizers modify the adhesive film so that the deformability and strength are improved. Among the most common plasticizers are urea, castor oil, glycerol, sorbitol, sodium acetate, sodium nitrate, and alkali thiocyanates.

The proportion of plasticizers added is adjusted so that no crystallization or exudation by synerisis occurs. The plasticizers function in 3 ways according to Kirkpatric: (1) by solvent action of the adhesive material, (2) by simple lubrication of the molecules layers, and (3) by a combination of these actions.

A common difficulty with dry adhesives is the formation of lumps on addition of water and to overcome this, Jagenberg-Wake has suggested treatment of the powder with a small amount of polyhydric alcohol, preferably an aliphatic glycol like ethylene glycol followed by heating to 80°C. This powder dissolves easily and the quantity of reagent required is only about 1%. If inorganic fillers are to be used in adhesives, they should be pasted with water containing a little soap to prevent lump formation.

Urea is an important additive, which functions in a number of ways. It reduces the viscosity of pastes, so that solid contents may be increased. This helps in preventing the curling or splitting of paper, which has been stretched during application of the adhesive while drying.

Urea also is considered to increase the tensile strength of the joint in addition to lowering the settling or gel point of the adhesive paste. Urea functions as a stabilizing agent by preventing setback of the adhesive, and this helps in reducing the degree of oxidation or the amount of alkali and processing required for achieving a particular fluidity. Urea also retards initial evaporation of water from the glue. This property is advantageous in glues for plywood, bentwood, and veneers. Urea and thiourea can solubilize the insoluble products obtained in the starch formaldehyde reaction. In a patent by Bauer, 5% urea and 5% alkal metal acetates have been used for remoistening adhesives.

The addition of soap provides thick pastes with smooth working properties for paperwork. A 10 to 12% starch paste containing 0.5% caustic liquor (36° Be′) gives a good paste on the addition of 1.2 to 1.5% soap before heating.

Detergents help in wetting foils used as wrappers, waxed paper, and highly calendered surfaces, which otherwise are very difficult to be wetted with the adhesive. Various surfactants used include sodium cetyl sulfate and disodium oleyl sulfate. A 2% addition of soap gives the desirable properties.

The addition of solvents for grease and waxes helps in the pasting of waxed papers, etc.

Some other interesting components added are decomposed grain flour, potato pulp, and partially fermented gluten.

Borates may also be used to increase stability and smooth working properties. Thus a barium starch may be kept mixed with borax in a dry form. When this mixture is added to water, an adhesive paste is obtained.

Clays and bentonites can be added to adhesives as fillers in order to reduce cost. They increase the solids content and inhibit penetration of the adhesive into the substrate. Bleaching agents like sodium bisulfite, hydrogen perioxide, and sodium perborate help in reducing film color.

Shelf life of adhesives is another important factor to be taken into consideration. The shelf life can be affected by microorganisms and hence antifungal agents may have to be used.

Among the modified starches, starch acetate is used in gummed tape formulation because of its flexibility, high gloss and rewetting ability. A gummed tape containing hypochlorite oxidized starch acetate with 1.5 to 2% acetyl groups and 0.3 to 0.5% carboxyl groups is reported to be similar in quality to that made with animal glue.[101] Hydroxylethylated starches are used in adhesives such as bag pastes, case sealing, and label and envelope adhesives. The characteristics which make it desirable are the water holding and filming properties. It is also used for the heavier usage of corrugative adhesives.[102]

Starch monophosphates blended with native starch, borax, sodium hydroxide, and water are used as adhesives in corrugated paper board and they have improved storage stability. Adhesives with good rapid bond strength are produced by combining neoprene rubber latex with 0.01 to 1.0% P starch phosphates. A latex containing about 4.2% starch phosphate when used to bond wood pieces gave tremendous increase in shear strengths.[103] Dialdehyde starch has also been used to produce water resistant adhesives.

D. Starch in the Dextrin Industry

Dextrin is a term used to indicate the degradation products of starch except mono- and oligosacharides in the broadest sense. Dextrins belong to a large varied group of D-glucose polymers which may contain purely linear, partially branched, highly branched, or cyclic compounds. Evans and Wurzburg[104] have classified dextrins into four major groups based on the general procedure used for the preparation:

1. Dextrins obtained by enzymic action, mainly amylases
2. Schardinger dextrins produced by action of the *Bacillus macerans*
3. Dextrins obtained by acid hydrolysis in aqueous media
4. Dextrins obtained by action of heat, or heat and acid on starch and termed pyrodextrins

Pyrodextrins are the most common dextrins which are widely manufactured and used.

Though raw starch possesses a high adhesive strength and provides the strongest films, it lacks some characteristics essential in the coating, binding, and adhesive application. In these applications, dextrins are preferred. Their relatively lower viscosities permit their use at a much higher concentration which make their films dry faster and provide a much faster hold than native starch.[104] Similarly, the solution stability at high concentrations is much more superior to the raw starch, hence they are more easily formulated in adhesives and the resulting adhesive has good working properties. Since the solubility of dextrins in cold water is higher, it can be used in applications which require partially or totally soluble binders.

The discovery of dextrins took place accidentally, when the starch stored in a textile factory in Dublin, Ireland, was partially burnt yielding a brown, sticky powder which was found to possess excellent adhesive power. Later it was found that dextrins having a wide range of solution properties could be obtained by dry roasting starch, either in the presence or absence of acid or alkali.

Much interest had been evinced on the mechanism of the dextrinization of starch. The first step is considered to be the hydrolysis of starch to relatively low levels of molecular weight followed by a recombination of the fragments primarily through (1-6) linkages to yield highly branched structures. During the initial stage, viscosity falls to very low levels. As the temperature rises, the repolymerization reaction becomes more pronounced and an

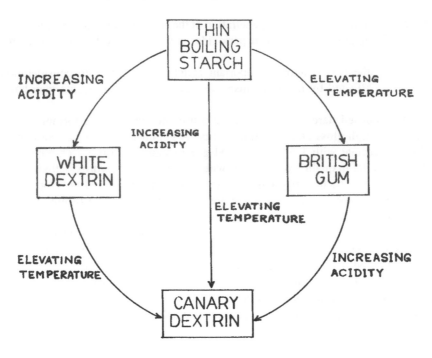

FIGURE 5.   Relationship between types of dextrins.

equilibrium in viscosity is reached. At still higher temperatures, trans-glucozidation becomes more important.

Dextrin appears in the market in varying forms — as powders of color varying from white to yellow to brown, as granulated particles, as a thick viscous-colored paste, or as a white paste.

The dextrins are generally classified as white dextrins, canary dextrins, and British gums. The products are obtained under different conditions and have widely varying properties and applications. The conditions for the preparations of the different types of dextrins have been standardized;[105] (Figure 5).

The white dextrins possess a white to cream color, possess variable viscosity, and solubility. Canary dextrins on the other hand, are buff to dark tan in color, have low viscosity in contrast to white dextrins and possess a high solubility. British gums can have color ranging form light to dark tan, and they possess variable viscosities and solubilities.

1. Types of Dextrins

a. White Dextrins

The preparation of white dextrins is carried out at low temperatures in the presence of the moisture in the starch and the catalyst. White dextrin varying from thin boiling starches to products with low viscosities are obtained by manipulating the amount of acid used. The solubility in cold water can be anywhere between 0 to 90%. White dextrins are widely heterogeneous, containing low molecular weight oligosaccharides and D-glucose. The average degree of branching in these dextrins is calculated to be around 3%.

b. Canary Dextrins

Canary dextrins are produced when the temperatures used are higher and the product assumes a tan color and increased solubility. When dissolved in water, they have a gummy appearance.

During the processing, a large degree of branching occurs due to repolymerization and transglucosidation.[106] The average degree of branching rises to 20% compared to 2 to 3% for white dextrins. Canary dextrin prepared from cassava starch has been analyzed to contain a high percentage of intermediate molecular weight fractions. Both high and low molecular weight fractions had disappeared from the mixture. In contrast, the cornstarch canary dextrin retains some of the high molecular weight fraction which is believed to be associated with the fatty acids present and is considered responsible for the thixotropic character of these dextrins.[104] Viscosity also varies with the acidity and temperature used for the conversion. A number of starches yield canary dextrins possessing high solution stabilities at solid concentrations as high as 70%.

c. British Gums

British gums are produced by heating untreated or buffered starch to high temperatures. Since no acid catalysts are used, hydrolytic breakdown is minimum and transglucosidation reaction is too slow. Hence longer periods of treatment are necessary to bring about the necessary rearrangement of the molecules of the product. Brimhall[107] considers the high molecular weight resulting from the treatment to be highly branched glucans similar to canary dextrins. The degree of branching is also in the same range, viz. 20 to 25%.

The solutions obtained on cooking British gums are more stable compared to the ones obtained from white dextrins of similar viscosities. Their higher molecular weight makes them setback more than white dextrins.

Among the different starches used for dextrin production, potato and cassava starches are the most preferred ones in view of the fact that they are the easiest to convert to dextrins. Corn and rice starches take a much longer time and a higher temperature to reach the same level of conversion. In fact, for top-quality work, cassava starch is the most ideal, because it renders a dextrin which has a slightly greater strength than potato dextrin and is odorless and tasteless so that it can be used in adhesives for postage stamps, envelope flaps, and labels, which are often wetted with the tongue, while potato dextrins possess a bitter taste.[108]

A combination of corn and cassava dextrin has been found to be ideal as a photographic mounting paste.[108]

It is necessary that the starch should be of the finest quality if a good lustrous dextrin is required. This is all the more important if the dextrin is to be used in paper making where black specks or discoloration is undersirable. In the case of cassava starch, the presence of HCN has been reported to slow down the rate of conversion of starch to dextrin.[110] The presence of sulfur compounds, which might have been added to improve color, is also not desirable.

2. Physical and Chemical Changes During Dextrinization

A number of changes — physical and chemical — take place during dextrinization. The main chemical changes are molecular size reduction and change in degree of linearity. The former affects the viscosity of the dextrin, while the latter influences the solution stability characteristics. The major changes and their nature are given below.

a. Moisture

The starch originally contains between 10 to 20% moisture and this is brought down to as low as 3 to 5% for white dextrin and less than 2% for canary dextrin. Since they are quite hydroscopic, if stored in a humid atmosphere, the values rise to 8 to 10%.

b. Viscosity

During the process of dextrinization, there is a rapid drop in viscosity during the initial stages, and this is more pronounced during the first hour of heating.[111] At high acidities, the rate of hydrolysis is a function of acidity. However after some time, an equilibrium is attained and no further fall in viscosity occurs. The simultaneous reactions — hydrolysis and repolymerization — may be shifted towards repolymerization and hence higher viscosities at higher temperature. The viscosity of dextrins can be measured using pipettes or in instruments which can measure the viscosity in centipoises. A special funnel has been suggested by Fetzer and co-workers.[107] The viscosity is a property which may be used to characterize a dextrin.

c. Solubility

During the early part of dextrinization, solubility is hardly increased, but at temperatures of 130 to 145°C under acidic conditions, solubility in cold water increases tremendously and reaches almost 100%. These products make unstable pastes in cooking. The increased solubility is brought about by the shortening of the chain length of the starch and the weakening of the hydrogen bonds which hold the granule together. Canary dextrin possesses high solubility and British gums show variable solubility from a few percent to almost 100%.

d. Reducing Sugar Content

The dextrose equivalent (DE) of starch during the initial phase increases and reaches a maximum at the point where viscosity shows minimum value. The formation of large quantities of saccharides including glucose, maltose, and oligosachrides imparts a high reducing value to white dextrin.[104] Further heating leads to repolymerization and also lowers reducing values and reducing sugar contents. The final sugar content of a canary dextrin is as low as 1%.

e. Specific Rotation

The specific rotation of the product at the early stage remains high and lies between +180° to +195°, but at the later stage, repolymerization and transglucosidation occur predominantly, and the value falls to nearly +150°. The drop in specific rotation indicates the formation of many new β-D-bonds which have inherently lower rotation.

f. Solution Properties

Solution stability varies over a wide range. This property is desirable in the case of canary dextrins, since in many operations, stability under high-solid content under acidic conditions may be required. The stability also is based on the source of the starch used. Dextrins made from cassava and potato starch are more stable compared to maize starch dextrin.[104]

g. Action of β-Amylases

White dextrins undergo almost a similar degree of conversion to maltose as starch does by action of β-amylase. The canary dextrin and British gums, on the other hand, give only a 12 to 15% conversion, again a reflection of the branched nature of these products.

3. Uses of Dextrins

An aqueous solution of dextrins can be used to form films capable of bonding similar or dissimilar surfaces and though these films are not as strong as starch films, they have a wide ranging application. They can be used at a higher concentration than starch and hence dry faster and provide a better bond. In addition, their increased solubility is an asset.

The adhesive industry is the major consumer of dextrins. They are used as back-steam and front-steam gums for envelopes, as bottle labeling adhesives, as adhesives in remoistening gummed tapes, postage stamps, lined cardboard boxes, and photographic mounting materials.

For the preparation of adhesives the dextrins are cooked in water at 95°C in upright cylindrical or horizontal trough-like tanks. These tanks are equipped with a propeller, anchor, or ribbon-type agitator and heating attachments. After pasting, the adhesive is packed for marketing.

A large number of modifiers are added to impart the desired characteristics to the adhesive. The additives modify the solution or film properties of the dextrins. Bonding agents increase the hold. The most common chemicals used are borax, caustic soda, and sodium metaborate. In addition, they increase the viscosity of the dextrin solution, stabilize the solution, and improve the paste color. The amount of borax added may be up to 20% of the weight of the dextrin and the chemical is added prior to cooking to achieve maximum efficiency. Addition of borax also reduces the concentration at which the dextrin forms a hydrogel. It enhances the rate at which the dextrin solution develops the bond. The effects of borax are most pronounced at lower levels of addition, and, by addition of alkali, the effect may be further enhanced.[105] Sodium hydroxide is usually added after cooking to reduce loss of viscosity which may occur by alkaline degradation of starch and to suppress color buildup. The action of sodium hydroxide is probably by conversion of borax to sodium metaborate. The addition of sodium hydroxide further increases viscosity, stability, and bond, while all the borax is converted to metaborate. It is also reported to enhance the bite of dextrin adhesives.[104] Levels above 0.5% sodium hydroxide, however, give unmanageably high cohesiveness.

Variuos chemicals are added to plasticize the dextrin film and reduce its tendency to become brittle at low levels of humidity. Three types of chemicals are effective in this respect — those which form a solid solution with the dextrins, humectants which control the moisture level in the film, and chemicals that lubricate the film.[105] The additives which belong to the first group are urea, sodium nitrate, dicyandiamide, salicylic acid, thiocyanates, formaldehyde, iodides, and guanidinium salts. They stabilize the dextrin solution against setback and also reduce the viscosity of the dextrin. Among these additives, urea is most commonly used and the quantity added is 1 to 10%. Glycerol, ethylene glycol, D-glucose, sorbitol, and many other polyhydroxy compounds act as humectants. They act by making the film dry slowly so that it will not become brittle, and will have desired bond flexibility. Usually sugars are used for this purpose, but glycerol and glycol are better humectants and they do not cause darkening of the bond with age.

The lubricants used include sulfonated castor oil, alcohols, and soluble soaps. They provide permanent flexibility at all atmospheric conditions, but the quantity used must be controlled to prevent excessive weakening of the adhesive bond.

Clays and bentonites are added as inert fillers to reduce cost. Bleaching agents such as sodium bisulfite, hydrogen peroxide, sodium peroxide, and sodium permanganate are used with high soluble dextrins in adhesives whose color is critical. Titanium dioxide may also be used to improve the whiteness.

Solvents like toluene, carbon tetrachloride, and trichloroethylene may be incorporated to increase adhesion to wax-treated and wax-impregnated paper. These solvents are added to cooked pastes after cooling. Preservatives such as formaldehyde or chlorinated hydrocarbons are used to prevent microbial growth.

In applications where water resistance is desired, insolubilizing agents may be used. Thermosetting resins such as urea-formaldehyde and resorcinol-formaldehyde impart the greatest level of water resistance. Thermoplastic resins, such as poly(vinyl acetate) and acrylics blend freely with starch paste and provide acceptable water resistance and machinability.[105]

Defoamers can be used to reduce the foaming during cooking and also during the use of the adhesives.

Some of the adhesive applications and typical formulations are discussed below.

a. Case and Carton Sealing

Case sealing involves sealing the top and bottom flaps of fiberboard and corrugated boxes and carton sealing is closing ends of smaller boxes usually made of fiberboard. They require adhesives having a fast bond and good machinability for high speed equipment. Borated dextrins with or without modification by caustic soda are mostly used for this purpose. The adhesives contain 30 to 40% solids and have a viscosity of 1000 to 3000 cP at 25°C. The formula consists of high-soluble white dextrin, sodium metaborate, and borax in water. When water resistance is desired, poly (vinyl alcohol) may be incorporated.

b. Laminating

Laminating may be of a general nature which involves paper-paper bonding or paper-paperboard bonding. The necessary criteria for a suitable adhesive for this purpose are a low adhesive penetration, a nonslipping nature, fast bonding and a noncurl characteristic. Plasticized white dextrin is quite suitable where a noncurling nature is desired. A formula given by Kennedy and Fisher consists of high soluble dextrin, clay, urea, borax and water. The final viscosity of the paste is around 500 cP.

Another type of laminating is solid fiber laminating which consists of the bonding of a large number of paperboards to give multiple laminations. This requires water resistance, a noncurling property, and good adhesive quality. Various types of dextrins can be used for the purpose and 5 to 25% urea-formaldehyde resins, or PVA and clay may be incorporated.[97] When metal foils are laminated to paper or paperboard, an alkaline pH should be imparted to the adhesive to neutralize any oil present on the metal.[97]

c. Tube Winding

This process is the production of paper tubes, cans, cones, etc. Most of the applications require fast bonding and high solid contents. These adhesives may be applied at 50 to 60°C to allow higher solids content. Many dextrin formulations have been used for this purpose.

d. Bag Adhesives

Grocery and multi-wall paper bags are increasingly being used and a large quantity of adhesives are used for the purpose. The two types of adhesives viz. side-seam adhesive and bottom-paste adhesives are used. The former is slow drying, low solid, and nonpenetrating, producing a high, dry bonded strength. White dextrins of 10 to 20% solids in the viscosity range of 1000 to 4000 cP are used. In order to provide for water resistance, thermoplastic and thermosetting resins such as PVA, acrylate polymers, urea, melamine or resorcinol-formaldehyde may be used.[97] Salt is added in some formulae to improve viscosity stability, and drying speed.

Bottom paste adhesives have a higher viscosity compared to side-seam pastes. The paste has a high thixotropic character and hence will not run through the applicator tip onto the bags passing below. But on application of shear, it becomes fluid and can be easily applied to the paper by applicator rolls. The bottom paste should necessarily possess good viscosity stability and hold to prevent the bottoms from opening before being bundled and wrapped. It should also be short and should provide strong bonding.

A third type of adhesive used in many multi-wall bags is cross paste. It glues various plies together before the tube is formed and seam paste applied. The cross paste is similar to seam pastes, but it should not penetrate through the outer layers. This has been achieved by the addition of poly (vinyl acetate) latex to the regular water resistant seam adhesive.[105]

e. Library Pastes

This paste is characterized by high viscosity, smooth texture, whiteness, and a neutral flavor. The formula consists of a mixture of starch, white dextrin, plasticizers (glycerol), deodorizer, and water.

f. Bottle Label Adhesives

The adhesive should have a good adhesion to glass, possess high viscosity and heavy body. The viscosity must be stable and the paste short and nonstringy. The solid content is 40 to 50%, and viscosity in the 80,000 to 150,000 cP range. Beer bottle labeling adhesive should withstand several days of immersion in ice water and should also be soluble in hot water so that the labels can be removed prior to reuse of the bottles.

g. Envelope Adhesives

The envelope adhesives are of two types — a back gum adhesive to hold the envelope together and front seal adhesive that is moistened by the user prior to sealing. The back gum adhesive has high solids to prevent adhesive penetration and to provide a good noncurling effect. A light and stable color, which does not darken on aging is preferred. The solids content is in the range 60 to 70% and viscosity 2000 to 5000 cP. A heavily plasticized dextrin or pregelled starch is used for the purpose.

The front seal adhesive should necessarily have a rewetting ability. The adhesive must have a stable viscosity, should not dry too fast, must have a noncurling effect, possess a high gloss and nonblocking character, and should not discolor the envelope. High-soluble dextrin from cassava, waxy corn, and potato are most suitable, but cassava dextrin with its bland taste is the most preferred one for this purpose.[108]

h. Flat Gumming

The adhesive for use in labels and trading stamps should be clear, light in color and glossy. The solids content is 50% and viscosity 1000 cP. The conventional process consists of applying at high temperatures a cooked paste of a highly soluble white or canary dextrin to the paper stock. However, curling is usually encountered. In the modified solvent process, a high soluble dextrin is suspended in an organic solvent along with a synthetic binder. The slurry has about 40% solids and viscosity of about 1000 cP. The slurry is coated on the substrate and the solvent evaporates leaving a discontinuous film. This discontinuity in the film prevents curling of the paper.

i. Gummed Tape

The adhesive used for reinforced sealing tape (which is required for applications where greater bond strength is required) must have a greater hold and is made using animal glue extended with dextrins. A patent for this adhesive uses waxy starch, dextrin, poly(acrylamide), dispersing agents, and water. The dextrin regulates the open time and the poly (acrylamide) improves the hold.

Crystal gum is a type of dextrin used for delicate work. It is preferred because it is neutral in reaction, is free from starch, contains very little sugar, and no chlorine or sulfur which usually impart color. Some of its applications are lining boxes with colored paper, to make the wood filler in cabinet-making, and in decorative work.[108]

An adhesive made by mixing casein, crystal gum, and hydrated lime or borax is used as a fixative for dentures, because it is tasteless, odorless, and resistant to saliva. The adhesive property lasts for sometime and the alkali serves the dual functions of assisting the casein to dissolve and slowing down the rate of hydrolysis of the crystal gum, which itself has a slower rate of hydrolysis compared to ordinary dextrins.

In addition to use in the adhesive industry, dextrins are used for various other purposes. Pyrodextrins find use in paper-coating operations. They are used in high solids machine coating in which the low viscosity and high binding strength are useful.[110]

Dextrins are used as sizers for glass fiber as soon as they are extruded through a platinum orifice. A film is formed by the dextrin around the fiber to protect it against abrasion.[104]

White dextrins are used in textile finishing operations in place of starch. Dextrins and

British gums are used as binders in various industries. They function as core binders in foundry operations, as binders for water colors, and for mineral aggregates, insecticides, etc.

Dextrins and British gums are used as thickeners for textile printing pastes which are made of an aqueous solution of chemicals, pigments, and dyes. British gum, in addition to thickening the paste, modifies the rheological properties so that it will flow freely onto the printing roll, engravings, or through printing screens, while preventing flow of the paste to the cloth. Dextrins with cold water solubility are used as carriers for active ingredients like food flavors. They are used as carriers and diluents for dyestuffs.[104]

Pharmaceutical companies also use white dextrins in the fermentation processes. The dextrin acts as a carbohydrate nutrient source in those instances requiring a slowly assimilable polysaccharide and not a rapidly consumed carbohydrate such as glucose. Johnson et al.[111] have patented this use in a Neomycin fermentation. The dextrins are often advantageous over starch since the high viscosity of the latter interferes with the agitation and aeration of the nutrient medium.

Thus, the role of dextrins increases with varying uses and newer avenues which are being continuously discovered. It is expected that they will have a steadily increasing demand.

E. Starch in the Food Industry

The acceptability and palatability of a food by a consumer depends to a large extent on the texture of the food, and starch contributes to the texture of many foods. The texture which has been defined as "appearance, feel to the touch, softness, and finally mouth feel" contributes to the concept of texture in a broad sense. Starch has varying functions in different products and the list of products in which starch is used is quite large. It provides cheap energy, body, and consistency to various products. The starch may be incorporated in its native form or a gelatinized form. Sago is an excellent example of a gelatinized food product from starch. Starch also finds use in various other foods and has diverse and far-ranging functions. Starch functions as an adhesive in breaded products. It is a binder in processed meats and extruded foods. The starch imparts cloudiness to cream fillings and sodas and glazing to nuts. It acts as a flowing aid in special sugars and baking powder, and is the foam-strengthening agent in marshmallows and sodas. In baked foods, starch provides antistaling character, while it helps in the gelling of gum drops and yieldings. It is also useful in the molding of gum drops, shaping meat products and pet foods, in stabilizing beverages and salad dressings, and in thickening gravies, pie fillings, and soups.[112]

The textural qualities are derived from the starch as a result of changes during and after cooking. Various factors which play a part in determining the starch texture are (1) gelatinization of starch, (2) swelling and viscosity, (3) paste clarity, and (4) gel formation and retrogradation.

Gelatinization is the phenomenon occurring when a starch slurry is heated and the temperature range over which the granules gelatinize completely is the gelatinization temperature. In addition to the type of starch, the presence of other materials, along with the starch, also affects the gelatinization temperature.

When gelatinization of starch occurs, it is accompanied by an increase in viscosity. This viscosity is desirable in products like cream soups, gravies, sauces, pie fillings, and puddings.[52]

However, often the paste acquires a stringy, mucilagenous character — not desirable in food products except in certain oriental dishes, in which the viscoelasticity is characteristic of the food. It is possible to improve the properties by special treatment or modifications. The viscosity is also modified by the presence of other components. The actual processing done in the kitchen may also be different from what is observed using a viscograph and hence allowances must be made for these effects. Gelatinization also increases the susceptibility of the starch to amylase degradation.

Whereas a starch suspension in water is opaque, the gelatinization increases the transparency and the pastes from different types of starch, possess varying degrees of transparency. The clarity is much higher for starch of waxy grains, root, and tuber crops compared to cereal starches. Clarity is also affected by the presence of other ingredients, as explained earlier. For example, sugars increase the clarity of cereal starches, whereas emulsifiers such as GMS make the paste more opaque. Sodium lauryl sulfate increases the clarity. Clarity contributes to the eye appeal of the food. In cherry pies and berry pies, filling appear more tempting when the fruit is clearly visible through the transparent juice than when concealed by a cloudy paste. Similarly, certain gravies of oriental foods possess higher clarity compared to those of many occidental gravies.

Starch solution forms a network upon cooling and when the gel is allowed to stand, water separates from the gel. Though gel formation is desirable for food products, retrogradation is not preferable, for example, the separation or "curdling" of frozen, starch-thickened sauces on thawing, and the skin on the surface of starch pastes.[52] This is more troublesome, when the food is subjected to repeated freezing and thawing operations. The stability can be increased by cross-linking and derivatization, and these modified starches have been successfully used in many food products.[54]

A comparison of different starches in relation to these properties has been detailed in earlier sections. It is seen that cassava starch has a long cohesive texture, with high paste clarity and good stability. With various types of modification, a starch of desirable quality can be obtained.

Cassava starch with its bland flavor is used in various food products in its pregelatinized form, e.g., pudding mixes, pie fillings, etc. Starch "pearls" are examples of partially gelatinized starch products. These are obtained by spreading damp starch on iron plates; it is stirred while being heated to cause partial gelatinization of the granules, which then agglomerate into pellets, and become hard and translucent on cooling. The long cohesive nature of the paste is prevented from developing in the products of pearl or "instant" precooked starch by keeping the agitation during cooking to a minimum. Pearl tapioca has been used for years to thicken various puddings. Along with banana starch, it has been widely used in various baby food formulas. It can be used in the production of custard powder, and along with other flours in making ice cream cones.

In addition to native starch, modified starch can also be used in food products. Acid-modified starches possess lower viscosity with higher gel strengths and better paste clarity. These acid-modified starches can be used for gum confections and cakes. Oxidized starches obtained by mild oxidation using alkaline hypochlorite provide gels of very low strength, but with improved clarity and are suitable for candy manufacture. Cross-linking reduces the stringiness without interfering with desirable properties. In addition, the cross-linked starches are resistant to acids. The common cross-linked starches are those made using phosphorous oxychloride, epichlorohydrin, metaphosphates, acrolein, cyanuric chloride, etc. The cross-linked root starches have been found to be suitable for use in products such as fruit pie fillings, canned pie fillings, etc., in view of the clarity and stability. In salad dressings, they are useful because of their resistance to acid hydrolysis. In addition, if cross-linking is combined with derivatization, the stability can be considerably increased, especially during repeated freezing and thawing operations.[54] Retrogradation may also be retarded by substituting some of the free hydroxyl groups with acyl groups, or by introducing ionizing groups which bring about repulsion between the molecules thereby preventing settling. The effect of ionizing groups has been made use of to increase stability of frozen white sauce thickened with carboxymethyl starch. The most important and promising derivative is the starch phosphate, in which part of the phosphate groups exists as a monoester not involved in cross-linking. These esters are highly stable and do not retrograde upon subjection to repeated cycles of freezing and thawing.[113] In view of their highly polyelectrolytic character, their clarity and waterbinding capacity are high. By manipulating the degree of substitution, the viscosity, solubility in cold water, and texture of the paste can be adjusted. Starch phosphate monoesters are good emulsifiers for vegetable oil in water. They are also used as pudding starches.

Various methods have been described to produce these esters. In one process, an alkali metal orthophosphate is used to yield a cold-water dispersible product. The solution of the product can be purified by treating with aqueous alcohol and the purified product has a higher viscosity and has tremendous stability towards freezing and thawing, both in a water paste and in a white sauce formula. It can be used as a substitute for vegetable gums.

The starch is suspended in water containing dissolved phosphate salts, the mixture is stirred for half an hour, filtered, dried at 40 to 45°C to 5 to 10% moisture, and then heat reacted to give a product of up to 0.2 DS. On a large scale, an intimate blend of 10% moist starch and orthophosphate at pH 5 to 6.5 is heated at 120 to 160°C for 30 min to 6 hr.

Meta-, polymeta- or pyro-tripolyphosphate, or a mixture of these, may also be used to prepare phosphate esters. By manipulating the amount of reagent used, the pH and the time of reaction, mono-starch phosphate and cross-bonded distarch phosphate esters can be formed simultaneously or in sequence in any desired ratio. Usually, waxy and tuber starches are used and their clarity and water absorption capacity are further improved. By allowing a small degree of cross-linking, the paste can be made to be nonstringy, short, stable to temperature, shear, and acid. Thus the product has all the desirable properties such as short texture, high clarity, and stability to repeated freezing and thawing. This product has been widely accepted in many types of starch-thickened frozen foods. Starches partially phosphorylated by spray drying with 0.5 to 1% phosphate, salts, followed by heat treatment, have been found to possess improved taste.[114] For use in food, the U.S. Food and Drug Administration (FDA) has permitted only the use of monosodium orthophosphate or sodium tripolyphosphate and sodium trimetaphosphate for esterification of starch and the residual phosphate in starch should not exceed 0.4% calculated as phosphorus.

Starch esters with organic acids also have special properties useful for food purposes. The esters of commercial interest are those which provide sol stability and hydrophobic, cationic, or anionic character at an economical level. Starch acetates and the half esters of some dibasic carboxylic acids have been found to be satisfying these conditions. Thus, acetylated starch prevents or reduces cloudiness and syneresis in aqueous dispersions of starch stored at low temperatures. Even as low a level as 5% substitution can bring about this stabilizing effect.[54] A number of acetylation reactions have been used for production of starch acetates. The reagents used in acetylation include acetic acid, acetic anhydride, vinyl acetate, acetyl phosphate, N-acetyl-N'-methylimidazodium chloride.[115] The native starch granule has little tendency to react with these acids or their derivatives, hence some pretreatment to breakdown the associative forces may be required. Among the different reagents, acetic anhydride is most commonly used. Using different solvents, catalysts, etc., various levels of substitution have been achieved. In use of sodium hydroxide solution as solvent and alkali, the sodium acetate formed as a byproduct has been found to cause off-flavors and odors in puddings made from starch acetates and, therefore, the starch acetate must be thoroughly washed to remove byproducts and maintain an ash content below 0.2%. The commercial process in the U.S. also makes use of acetic anhydride in the presence of dilute alkali.

The major advantage of acetylated starch is the clarity and stability of the paste even at low temperatures, not found with native starches, including cassava. In combination with cross-linking, the acetyl starch is used in canned, frozen, baked, and dry foods, and also in baby foods. They are used in fruit and cream pie fillings in cans and jars which require storage under varying temperatures. Frozen fruit pies, pot pies, and gravies are other examples in which acetylated starch is used. Syneresis is reduced in baked goods on use of acetylated starch. In dry mixes, instant gravies, and pie fillings, pregelatinized starch acetate is used. The Food and Drug Administration (U.S.) has permitted up to 2.5% acetyl content in starches used in foods.

Hydroxyalkyl starches like acetylated starches have good dispersion stability (DS) and nonionic character. The amount of this ether produced annually is approximately 90,000 tons. The production involves treatment with alkylene oxide in the presence of alkali. Catalysts may also be used to provide various DS products with different properties. Low DS hydroxyethyl and hydroxypropyl starches behave like low substituted starch acetates. The pancreatin digestibility is, however, reduced, especially with gelatinised starch derivatives. These derivatives also are useful in food application because they provide viscosity stability and water-holding capacity under low temperatures. They have also been used along with other thickeners, e.g., with carrageenan in the milk system and with xanthan gum in salad dressings. They are also used in gravies, sauces, pie-fillings, and puddings because of the smooth, clear, thick nongranular texture even after freezing. Hydroxypropyl cassava starch has been found to be very suitable for frozen puddings.[116]

Slightly oxidized starches have been found useful in batters and breaded coating for foods such as fried fish, as the starch offers a good coating to the food.[117] In addition, the heat moisture treated starch may be used to improve properties of root starches with high swelling tendencies.

Pregelatinized starches, as mentioned earlier, are receiving much attention, because of their solubility, digestibility, and ease of preparation. The most popular food based on pregelatinized starch is "instant pudding", which is available in packed powder, which, when mixed thoroughly with milk and allowed to swell for 5 min, yields simple puddings. The formula consists of pregelatinized starch with sugar, flavorings, and suspending agents (mostly salts) which produce enough viscosity in the milk to keep the starch suspended until completely hydrated. It is found that cassava starch is the most ideal for this purpose and Minute Rice® is a long-established brand sold as a dessert item in the U.S. by General Foods Corporation and by Tipiak in France. It is also used as an ingredient in baby foods and marketed by Gerber® and in the Jello® brand of desserts.[71] It has been estimated that the amount of imported cassava starch consumed in the U.S. in modified and unmodified form was 87% between 1980 and/81. In West Germany, Dr. Oetkar's group of companies uses unmodified starch in a number of products, including a packeted product for "tortenguss", a clear jelly topping for tarts.[71] "Amylum" and "Avebe" also use cassava starch to smaller extents. In the U.K., the major end use is for puddings, which include institutional sales to schools and hospitals. Baby foods are also made using cassava starch. In continental Europe, the tapioca pearl is more often used as a thickener in soups. The major portion of the tapioca starch used in European communities is for food use, accounting for about 90% of total imports. Of the starch imported, 50 to 60% is converted to modified products which are mostly used for instant puddings, baby foods, and confectionery.[71]

It is interesting to note that most of the countries which produce cassava starch do not make any of these food products themselves, except Thailand for the production of "Chinese style" foods. However, with the increased demand for instant and convenience foods in the developing countries, the demand for cassava starch for food products is inevitable. Hence there is a good future for cassava starch in the food industry, both in developed and developing nations.

F. Starch in the Sweetener Industry

Starch is a polymer made up of thousands of glucose units and hence breakdown of starch yields glucose as the final product. The discovery that heating starch with dilute sulfuric acid gives rise to a sweet substance was made by Kirchoff in 1811. Though originally this sweet substance was thought to be sugar, Saussure in 1814 proved that it was glucose.[118]

It is no exaggeration to say that nowadays, starch production is directly linked to glucose production and innumerable factories all over the world are turning out a large quantity of glucose in various forms. New methods to improve yield and quality at reduced cost are constantly being devised.

Commercial starch hydrolysates are classified according to the dextrose equivalent (DE) of the syrup.[119] Maltodextrins have a DE less than 20. Those above 20 DE may be further sub-classified depending on the content of dextrose. The syrup having DE 20 to 38 contains mainly high molecular weight branched and linear dextrins. High conversion syrups contain 75 to 85% D-glucose, maltose, and maltotroise. Each of these syrups has its own characteristic property and its own applications.

1. Maltodextrins

Maltodextrins are manufactured in a manner similar to glucose syrups, except that the hydrolysis level is controlled to have a DE of less than 20. The hydrolysis is catalyzed by acid, enzyme, or both. Though maltodextrins are usually prepared by acid hydrolysis, the product suffers from the development of haziness or cloudiness caused by retrogradation of the linear starch fragments. Acid-enzyme or dual-enzyme hydrolysis overcomes this problem by hydrolyzing the linear dextrins in preference to the branched dextrins. Armbruster and Hajres suggested a two-stage hydrolysis, first to a DE 5 to 15 by acid catalysis, followed by neutralization and further hydrolysis by a bacterial amylase. In another process, starch slurry is liquified by heating 70 to 90°C in the presence of α-amylase to DE 2 to 15. It is then autoclaved at 110 to 115°C to gelatinize the remaining insoluble starch and cooled and subjected to further enzymatic hydrolysis.

The hydrolyzed mass is brought to a pH of 4.5 and filtered. Filtration poses problems since viscosity is high and often loss of material occurs. The clarified solution is refined by standard procedures and then concentrated in a vacuum to syrups containing 75% solids or to white powders.

The properties of maltodextrins depend on the DE level achieved. They are relatively nonhygroscopic and the lower the DE, the less will be the tendency to absorb moisture. Maltodextrins generally possess high viscosity due to the presence of a large amount of high molecular weight saccharides.

2. Glucose Syrups

The production of glucose syrups is well documented and a number of patents have been filed for acid, enzyme, and acid-enzyme processes. The acid-catalyzed hydrolysis may be carried out in either a batch or continuous process.

In the batch process, the starch slurry and hydrochloric acid at a ratio of 0.12% acid to the weight of starch used, are heated in cylindrical stainless steel or copper vessels, at a temperature of 132 to 137°C using steam. When only a small quantity of dextrin remains, the pressure is released and the mass is transferred to a neutralization tank where the neutralization is carried out using an alakali, usually lime. It is filtered and concentrated to a density of 30° Bé. and filtered through animal charcoal to obtain a colorless syrup, which is further concentrated in a vacuum to 42 to 45° Bé. and then packed.[120]

In a typical continuous hydrolysis process, a 40% starch slurry is acidified to pH of nearly 2.0, and then pumped into the hydrolyzer. Simultaneously, steam is injected to heat the slurry to 140°C. After hydrolysis, the steps are the same as in batch hydrolysis.[121]

It is generally observed that for obtaining higher DE, acid hydrolysis is not suitable since it gives rise to side reactions which lead to excessive color, bitterness, etc; Gentiobiose is one such bitter disaccharide formed. The discovery of a dual acid-enzyme process in 1940 opened a new era in glucose syrup production. The enzyme preparation usually contains a mixture of β-amylases and glucoamylases and are obtained from *Aspergillus oryzae*. The steps involve the initial conversion of starch by acid to dextrose equivalent (DE) 44. It is then neutralized and concentrated and the enzyme added. The optimum conditions are pH 5.2 and temperature 55°C. After hydrolysis, the pH is adjusted with sodium carbonate and then filtered and centrifuged. It is concentrated to about 55% solids and refined using activated carbon. The carbon treatment effectively removes any 5-(hydroxymethyl)-2-furaldehyde formed during acid treatment. If necessary, it may be further purified by ion exchange treatment. The strength is increased to the desired level by vacuum evaporation and cooled and packed in air tight drums, since it is quite hygroscopic.

The sweetness of the syrup depends on the DE. It can be diluted in water and will retain its clarity, if packed and stored properly.

When high conversons are required, the use of enzymes becomes necessary. Acid catalyzed hydrolysis does not yield products with more than 90% D-glucose, because acid-catalyzed reversion and dehydration reactions become prominent, leading to a reduced quantity of D-glucose. Enzyme-catalyzed hydrolysis is optimally done at a pH of 4 to 5 and does not require high temperatures. Liquefaction of starch is first achieved by using α-amylase and the saccharification by glucoamylase. The α-amylase converts amylose and amylopectin to linear and branched dextrins, which are converted by glucoamylase to glucose. However, even under the best conditions, 100% D-glucose is not formed, since some degree of reversible reactions occurs leading to disaccharides.

Many factories have shifted to enzyme hydrolysis of starch because of the cleaner reaction, better conversion, and lower cost. But the enzymes must be produced cheaply and should be pure.

After hydrolysis, the refining is carried out as described for the acid-enzyme process. For production of dextrose, the ion-exchange resin treatment is not compulsory, whereas it is necessary for production of high fructose syrup, since the presence of calcium ions and other impurities can lead to inhibition and deactivation of glucose isomerase. Consequently, the decolorized syrup is passed through a bed of strong acid, cation exchange resin form in the acid, and then through a bed of weak base, anion exchange resin is produced in the free base form.

The final syrup is sweet, odorless, and colorless, and may be used for dextrose preparation or production of high fructose syrup.

Immobilization of amylases has been used to shorten the time of reaction and reduce side reactions.[122] The enzyme can be recovered and reused. However, so far, no commercial applications have been reported.

Similarly, immobilization of gluco-amylase has been tried. In one process, freshly prepared solutions of α-amylase hydrolyzed dextrins of DE 25 to 30 are fed through an immobilized gluco-amylase bed.[123] At 27% feed concentration, the product contained 94% glucose. But even this process has not been accepted commercially.

3. Crystalline Dextrose

Dextrose may be obtained in the monohydrate or anhydrous forms. The high conversion syrup is refined by passing through ion exchange resins to reduce color and ash content. It is concentrated under vacuum to a 70 to 78% solid content, cooled, and taken to a crystallizer which may be batch or continuously operated. The conditions for proper crystallization are control of the D-glucose content and dry substance in the crystallizer liquor. A particular level of supersaturation is required to obtain good sized crystals. Since the crystallization is exothermic, a cooling stage must be provided. At the end of crystallization, the mass is taken to large centrifuges lined with perforated screens in which the crystalline solids are separated from the mother liquor. The cake is washed with a small quantity of water, and the wet cake containing 14% moisture is dried to a 8.5% moisture content, slightly less than the 9.1% calculated for monohydrate crystals. A second crop of crystals is obtained which is as pure as the first crop. The resultant liquor after the second crystallization, called "second greens" or "hydrols", is usually sold as raw material for food products.

Anhydrous α-D-glucose is made by redissolving and recrystallizing α-D glucose monohydrate in a batch vacuum evaporator/crystallizer. Evaporation during crystallizaiton balances the supersaturation required. Seeding is done by initially concentrating a limited portion of D-glucose solution at 65° C until crystal nuclei of anhydrous α-dextrose form and these are then allowed to grow by controlled addition of a solution of D-glucose. The crystals obtained are centrifuged, washed with water, and dried to less than 0.1% moisture. The yield is around 50% and the liquor is used for glucose hydrate manufacture.

In addition to the above, "total sugars" obtained by dehydrating glucose syrups or fructose syrups are also finding varying uses.[119] They contain a mixture of monosaccharides and oligosaccharides, and provide the characteristic properties of some food products in which they are used.

4. Fructose Syrup

The remarkably higher sweetness of fructose compared to glucose had been recognized much earlier and the conversion of glucose to fructose had been a scientific dream. Base-catalyzed conversion was tried, with little success, by many workers. The problems encountered in hydroxide-ion catalyzed isomerizaton are production of mannose as a major byproduct, impurities such as psicose, and large amounts of degradation products leading to a high ash content and discoloration which make refining and separation difficult and uneconomical.[119] Hence, the development of an efficient enzymatic means for the conversion of glucose to fructose was a boon to the sweetener industry. The major breakthroughs in the process were the recognition by Marshall and Kooi[124] that D-xylose ketol isomerase catalyzes the interconversion of glucose and fructose, the discovery of certain *Streptomyces* sp. as rich source of heat-stable glucose isomerase, and more recently, the development of low cost and efficient methods for immobilization of glucose isomerase.[119]

Various organisms have been recognized to produce glucose isomerase, but only *Streptomyces*, *Bacillus*, *Arthrobacter*, and *Actinoplanes* yield thermostable isomerase.[125] The isomerases now commercially available are obtained from isolated and mutant cultures which do not require xylose as a substrate.

The enzyme is used almost exclusively in immobile form. The methods have been summarized by various workers, and the most feasible methods are (1) adsorption on an insoluble carrier, and (2) fixation of the enzyme within the cellular microorganism on which it occurs.[126] The former process involves extraction of the intracellular isomerase by disintegration followed by adsorption on an insoluble carrier like DEAE cellulose. The latter process consists of heating the harvested mycelia, binding the heat-treated intact cells to polymeric agents such as gelatin and chitosan, and reinforcing the binding by cross-linking it with glutaraldehyde.

The glucose syrup obtained from starch hydrolysis, after purification to remove all ions, is passed through the medium containing the isomerase. The optimal conditions for isomerization are (1) glucose concentration at 40 to 60%, (2) pH 7 to 8.5, and (3) temperature of 55 to 65°C. The calcium content in the glucose syrup should be below 20 ppm, or it will inhibit the enzyme. A small quantity of Mg^{++} ions is required as an enzyme activator. The contact time for conversion is less than 1 hr. These mild conditions ensure that no sugar degradation in the products are obtained unlike alkali-catalysed isomerisation.

The isomerized syrup is refined by carbon and/or ion exchange treatment and concentrated at low temperature under vacuum to a syrup containing enough solids to prevent microbiological spoilage, but low enough to prevent crystallization of glucose. The resulting syrup is clear, colorless, and has the following charactertistics: solid content 70 to 72%; pH 4.0; D-glucose 52%; fructose — 42%; other saccharides — 5%; and relative sweetness at 15% concentration (compared to sucrose = 100).

A syrup containing 90% fructose and 7% glucose may be prepared by various methods such as preferential complexation and chromatographic separation, and the sweetness is more than 42% fructose syrup.[119,127] In addition, crystalline fructose may also be obtained.

All these products have been widely accepted as sweeteners in confectioneries, colas, etc; and the demand for them is steadily increasing. In the U.S., the high-fructose corn syrup has almost completely replaced other sweeteners in colas and soft drinks. Many other countries have also developed high-fructose syrup for commercial use.

5. Uses of Sweeteners

Maltodextrins find use in a variety of foods. These include dry mixes such as soup mixes, fruit-flavored mixes, dairy drinks, cake, and cookie mixes. Maltodextrin is preferred over glucose syrup in many foods in view of its lower hygroscopicity, bland, nonsweet taste, and the ability to give a "bodying effect" due to its relatively higher viscosity. Starch provides viscosity, but its solubility is poor compared to maltodextrins. Newer uses are being found for maltodextrins which are assuming more importance.

Glucose syrup has long been used in certain industries, especially in the confectionery industries. Glucose is preferred over sucrose as a sweetening agent in many foods. A mixture of these two finds wide use in the manufacture of confections. The optimum concentration is 20 to 50% of total sugar, since above 50%, crystallization occurs. Canned sauces, tomato juice, sweet pickles, etc. are made using glucose syrup. Glucose syrup, in addition to adding sweetness, has other functions such as controlling humectancy and hygroscopicity, imparting texture, body, and cohesiveness, and controlling the crystallization of sugar.

Glucose syrup is also used in adhesives and in foundries. It is the base for the manufacture of such industrial chemicals as alcohol, gluconic acid, acetone, butane diol, acetic acid, fructose syrup, citric acid, and sorbitol. It is also used in pharmaceutical syrups.

Dextrose is used in the food industry in baking as it serves as a fermentable sugar, and also aids in the enhancement of flavor and aroma and contributes to crust color.[128] In the dairy industry, it is used in frozen desserts to control excess sweetness and for flavor improvement. Dextrose is also used in the production of tablets in the food and pharmaceutical industries. The cooling effect produced by the dissolution of dextrose monohydrate in the mouth enhances certain flavors. Dextrose also finds wide use in the pharmaceutical industry for intravenous feeding and formulae. In foundry industries, it serves as a binder in foundry cores because of its ability to polymerize under heat in the presence of a catalyst.

High fructose syrup, the revolutionary sweetener, has almost displaced glucose syrup in the soft drink industry. It is used in the production of carbonated and noncarbonated beverages. The syrup functions by producing sweetness at lower levels of incorporation and also balances the flavors and acids to give an acceptable product. Fructose syrup also finds use in the canning of fruits, and the manufacture of jams and jellies. The bakery industry also consumes fructose syrup. With the arrival of 90% fructose, which is 1.5 to 1.8 times sweeter than glucose the sweetener industry producing fructose syrup has a promising future.

Syrup solids obtained from evaporative drying of hydrolyzed starch syrup are widely used in the production of whiteners. Fructose in crystalline form is incorporated into dietetic foods because of its lower caloric values and also in beverage mixes, gelatin desserts, cakes, and cookies.

G. Miscellaneous Applications of Starch

In addition to the major industries which use starch, there are a number of other applications of starch which are encountered in many fields. Though these industries use starch in small quantities, the total quantity of starch utilized is quite large. Some of the major industries are described below.

1. Soap and Detergent Industry

Starch functions as a filler in soaps and detergents in maximum concentrations of 15%. The starch should have high gloss, maximum whiteness, and it should be free from chlorine and acid, and the maximum moisture content should be less than 20%. The starch is incorporated in powder form to the chips before milling.[129] Cassava starch can be conveniently used for the purpose, provided the viscosity and color are uniform.

2. Laundry Starches

Cassava starch, with its lower gelatinization temperature will undergo a higher degree of gelatinization in the washing machine, and hence can be preferentially used as laundry starch. Starch is used along with other ingredients like borax, china clay, and some fats. However, the stickiness of the starch solution is a problem commonly encountered.

3. Cosmetic Uses

Starch is widely used a diluent in many types of toilet powders.[129] The starch for this purpose should be perfectly dry. The quantity of starch used is reportedly as high as 50%, but its use depends on the price factor in relation to other fillers.

Preparation of dry shampoos for hair using modified starch and surfactants has been reported.

4. Pharmaceutical Uses

Starch has been widely used in making pills, in which it serves a dual role — as a coating and dusting agent and as a binder for the active ingredient in the pill. In aspirin tablets, the function of starch has been explained by Radley.[129] When the tablet is swallowed, the starch rapidly absorbs moisture and swells, causing an internal stress which disintegrates the tablet to release the active ingredient. Starch is also used in some insecticide powders. Iodine complexes of starch ethers have been used as disinfectants.

5. Horticultural Uses

Starch is often used in the preparation of horticultural sprays. The function of the starch is to help the dissolved material to adhere to the treated area, even after the liquid medium has evaporated.[129] Yellow dextrins may also be used.

6. Fire-Proofing Preparations

Various formulations including starch have been used in nonflammable fabrics. The starch, acting as a size, causes the other ingredients, which are used for fire proofing, to adhere to the fabric.

7. In Explosives

In various explosives, starch is incorporated, again as a filler capable of being inflamed. Various types of explosives using starch have been explained by Radley. Starch can also be used as a binder in match heads and in fireworks. The starch functions not only as a substitute for the more costly hide glue, but acts as thickener and binder, which is easily oxidized and aids in combustion. The amount of starch added is 13 to 14%.

8. In Drilling Muds

Oxidized starch is used as a dispersant in drilling muds.[130] It lends improved dispersability and high temperature stability in the drilling fluid. Starch fits very easily into the thixotropic colloid system of the drilling mud. The properties are further improved when borax is also incorporated. Many other additions have also been suggested.

9. In Optical Whiteners

A starch derivative made by starch reacting to a substituted cyanuric chloride was found to be an effective optical whitener.[131]

10. In Leather Treatment

Oxidized starch and sodium isophthallate produce a good chemical for leather treatment.

11. As Plasma Expander

Starch plasma expanders of predetermined intervascular persistence, obtained by etherification of controlled hydrolyzed starch have been patented by Burbank et al.[132]

12. In Production of Polyhydroxy Compounds

The production of sorbitol has already been mentioned. In addition, various other industrially important polyols have been made from starch. A polyol prepared from starch reacted with ethylene glycol has potential to be used in rigid urethane foams, surfactants, and alkyds.[133] The glycols can be used as initiators for making polyethylene for rigid urethane foam production. They possess biodegradability not found with synthetic polyurethanes.[134]

13. Plastics from Starch

The ever-increasing use of plastics has led scientists to think in terms of using nonpetroleum-based raw materials. In addition, the waste disposal problem of plastics has brought to light the utilization of biodegradable plastics. Starch has been evaluated as an inert filler in PVC plastics, and as a reactive filler in rigid urethane foam. The tensile strength was maintained even at a 50% level of incorporation of starch, clarity was good, but elongation was reduced. They exhibited good biodegradability.

Agricultural mulches made of starch-incorporated biodegradable plastics also provide potential application for starch. Starch poly-(vinyl alcohol) films may be used for this purpose. A water-soluble laundry bag for use as disposable bags in hospitals has been marketed in the U.S. since 1977. Another use is in packaging for chemical pesticides. It has been found that starch is quite compatible with synthetic polymers, even forming chemical bonds.[135] Urethane starch plastics contain starch bonded to a resin. A typical formula used is 10 to 60% starch, castor oil, polymeric diisocyanates, etc. Modification of starch can enhance the linkage.

14. Graft Copolymers of Starch

Graft polymerization can be used to bring about chemical bonding between natural and synthetic polymers. The process involves initiating a free radical on the starch backbone and then allowing the radical to react with polymerizable molecules. Both chemical and irradicated initiation of free radicals have been achieved. Some examples of these products are starch-graft-polystyrene, - poly (methyl methacrylate), poly (methyl acrylate), etc; these products using 20% starch are extrudable and have good tensile strength. These graft polymers have such varying applications, as thickeners for aqueous systems, flocculents, clarifying agents for wastewater, etc. An interesting product developed in the U.S. is a polymer "super slurper" made using starch and polyacrylonitrile. The product has the capacity to absorb many times its own weight of water and has wide applications, such as water retention in seed coatings, in bandages, disposable diapers, and incontinence pads.[136,137]

15. Starch Xanthide in Rubber

Use of starch as a replacement for carbon black in rubber processing has been found feasible. A new innovation is based on the encapsulation of a wide range of chemical pesticides in a starch matrix using starch xanthide. They have a longer shelf life, higher safety in handling, and act as a slow release system. Newer developments in these fields are showing good promise.[134]

Thus we find a variety of uses for starch in different fields ranging from food to medicine, effluent treatment to soil conditioning, biodegradable plastics to explosives, and a very large spectrum of other uses.

REFERENCES

1. **Schoch, T. J.**, The fractionation of starch, *Adv. Carbohyd. Chem.*, 1, 247, 1945.
2. **Meyer, K. H. and Bernfield, P.**, Research on starch, V-Amylopectin, *Helv. Chim. Acta.*, 23, 875, 1940.
3. **Zuber, M. S.**, Genetic Control of Starch Development, in *Starch: Chemistry and Technology*, Vol. 1, Whistler, R. L. and Paschall, E. F., Eds., Academic Press, New York, 1965, 4.
4. **Moorthy, S. N. and Maini, S. B.**, Varietal differences in cassava starch properties, *Proceedings of Seminar on Post Harvest Technology of Cassava*, AFST (I), Trivandrum, India, 1980, 58.
5. **Raja, K. C. M., Abraham, E., and Mathew, A. G.**, Effect of defaulting on amylose content, viscosity characteristics and organoleptic quality of cassava, (*Manihot esculenta* crantz), *J. Food Technol.*, 17, 761, 1982.
6. **Banks, W. and Greenwood, C. T.**, *Starch and its components*, Edinburgh University Press, Scotland, 1975, 247.
7. **Howarth, W. N., Hirst, E. L., and Isherwood, F. A.**, Polysaccharides: Part XXIII, Determination of Chain length of glycogen, *J. Chem. Soc.*, 577, 1937.
8. **Staudinger, H. and Husemann, E.**, Starch and its structure, *Ann. Chem.*, 27, 195, 1937.
9. **Gunja Smith, Z., Marshal, J. J., and Smith, E. E.**, Enzymatic determination of the unit chain length of glycogen and related polysaccharides, *FEBS Lett.*, 13, 309, 1971.
10. **Robin, J. P., Mercier, C., Charbonniere, R., and Quilbot, A.**, Lintnerised starches: gel filtration and enzymatic studies of insoluble residue from prolonged acid treatment of potato starch, *Cereal Chem.*, 51, 389, 1974.
11. **Robin, J. P., Mercier, C., Duprat, F., Charbonniere, R., and Guilbot, A.**, Lintnerised starches: chromatographic and enzymatic studies on the insoluble residues remaining after hydrolysis of cereal starches, especially waxy maize starch, *Starke*, 27, 36, 1975.
12. **Banks, W., Geddes, R., Greenwood, C. T., and Jones, I. G.**, Physico-chemical studies on starches: part 63, The molecular size and shape of amylopectin, *Starke*, 24, 245, 1972.
13. **Hood, L. F. and Mercier, C.**, Molecular structure of unmodified and chemically modified manioc starches, *Carbohyd. Res.*, 61, 537, 1978.
14. **French, D.**, Organisation of starch molecules, in *Starch Chemistry and Technology*, 2nd ed., Whistler, R. L., Bemiller, B. N., and Paschall, E. F., Eds., Academic Press, New York, 1984, chap. 7.
15. **Finkelstein, R. S. and Sarko, A.**, Anistotropic scattering by single starch granules: II. Layered granule structures, *Biopolymers*, 11, 881, 1972.
16. **Yagamuchi, M., Kainuma, K., and French, D.**, Electron microscopic observations of waxy maize starch, *J. Ultrastruct., Res.*, 69, 249, 1979.
17. **Kainuma, K. and French, D.**, Naegeli amylodextrin and its relation to starch granule structure: II. Role of water in crystallization of B starch, *Biopolymers*, 11, 2241, 1972.
18. **Laloir, L. F., de Fekete, M. A. R., and Cardini, C. E.**, Starch and oligosaccharide synthesis from uridine diphosphate glucose, *J. Biol. Chem.*, 236, 636, 1961.
19. **Radley, J. A.**, Physical methods of characterising starch, in *Examination and Analysis of starch*, Applied Science Publishers, London, 1976, 123.
20. **I. S. I.**, Specification for tapioca starch for use in the cotton textile industry, Indian Standard Institute, New Delhi, 1970, 1605-1960.
21. **Radley, J. A.**, Tapioca, cassava or Brazilian Arrow root starch, in *Starch Production Technology*, Applied Science Publishers, London, 1976, 206.
22. **Ripperton, J. C.**, Physicochemical properties of edible cassava and potato starches, *Hawaii Agr. Exp. Stn. Bull.*, 63, 1, 1931.
23. **Kerr, R. W.**, in *Chemistry and Industry of Starch*, Academic Press, New York, 1944, 109.
24. **Moss, G. E.**, Microscopy of starch, in *Examination and Analysis of Starch*, Radley, J. A., Ed., Applied Science Publishers, London, 1976, 18.
25. **Burre, J., Bonn, F., and Delarouzee, R.**, Cassava from Indochina, *Annales des Falsif. Fraudes*, 43, 129, 1950.

26. **Snyder, I. E.,** Industrial Microscopy of starches, in *Starch Chemistry and Technology,* Whistler, R. L., Bemiller, J. N., and Paschall, E. F., Eds., Academic Press, New York, 1984, chap. 22.
27. **Radley, J. A.,** *Starch and its Derivatives,* 2nd Ed., Applied Science Publishers, London, 1957, 3.
28. **Fitt, L. E. and Snyder, E. M.,** Photomicrographs of starches, in *Starch Chemistry and Technology,* Whistler, R. L., Bemiller, J. N., and Paschall, E. F., Eds., Academic Press, New York, 1984, chap. 23.
29. **Banks, W. and Greenwood, C. T.,** *Starch and its Components,* Edinburgh University Press, Scotland, 1975, 102.
30. **Hood, L. F.,** Current concepts of starch structure, in *Food Carbohydrates,* Linebeck, D. R. and Inglett, G. E., AVI Publishing Company, Westport, Conn., 1984, 217.
31. **French, D.,** Organisation of starch granules, in *Starch Chemistry and Technology,* Vol. 2, Whistler, R. L. and Paschall, E. F., Eds., Academic Press, New York, 1984, 7.
32. **Gallant, D. J., Bewa, H., Buy, Q. H., Bouchet, B., Szylit, O., and Sealy, L.,** On ultrastructural and nutritional aspects of some tropical tuber starches, *Starke,* 34, 255, 1982.
33. **Greenwood, C. T.,** Observations on the structure of the starch granule, in *Polysaccharides in Food,* Blanshard, J. M. V. and Mittchell, J. R., Eds., Butterworths, London, 1979, 8.
34. **Ceh, M., Stropnik, C., and Leskovar, S.,** Potentiometric determination of molecular weights of starches, *Starke,* 28, 51, 1976.,
35. **Schoch, T. J. and Jensen, C. C.,** *Industr. Eng., Chem., (Anal., Ed.)* 12, 531, 1940.
36. **Leach, M. W.,** Gelatinisation of starch, in *Starch Chemistry and Technology,* Vol. 1, Whistler, R. L. and Paschall, E. F., Eds., Academic Press, New York, 1965, chap. 12.
37. **Moorthy, S. N.,** Acetylation of cassava starch using perchloric acid catalysis, *Starke,* 1985.
38. **Moorthy, S. N.,** Acetylation of cassava starch, in *Proc. of Seminar on Post Harvest Technology of Cassava,* AFST (I), Trivandrum, India, 1980.
39. **Krog, N.,** Influence of food emulsifiers on pasting temperatures and viscosity of various starches, *Starke,* 25, 22, 1973.
40. **Moorthy, S. N.,** Effect of different types of surfactants on cassava starch properties, *J. Agric. Food Chem.,* 33, 1227, 1985.
41. **Moorthy, S. N.,** Behavior of cassava starch in various solvents, *Starke,* 32, 321, 1982.
42. **Oosten, B. J.,** Effect of organic molecules on the gelatinisation of starch, *Starke,* 36, 18, 1984.
43. **Gerisma, S. Y.,** Gelatinisation temperature of starch, as affected by polyhydric and monohydric alcohols, carboxylic acids and some other additives, *Starke,* 22, 3, 1970.
44. **Smith, R. J.,** Viscosity of starch pastes, in *Methods in Carbohydrate Chemistry,* Vol. 4, Whistler, R. L., Ed., Academic Press, New York, 1964, 120.
45. **Meiss, P. E., Treadway, R. H., and Smith, L. T.,** White potato starches, *Industr. Eng. Chem.,* 36, 159, 1944.
46. **Elder, A. L. and Schoch, T. J.,** A new viscometer for starch viscosity measurement, *Cereal Sci. Today,* 4, 202, 1959.
47. **Zoebel, H. F.,** *Starch Chemistry and Technology,* 2nd ed., Whistler, R. L., Bemiller, J. N., and Paschall, E. F., Eds., Academic Press, New York, 1984, 9.
48. **Brabender, C. W.,** The structure of wheat and rize doughs, *Muhlenlab,* 1, 121, 1937.
49. **Muller, G. J.,** The pasting characteristics of starches and their influence on the quality of cereals, flours, and other products containing starch, C. R., *VI Cong. Budapest,* 2, 529, 1939.
50. **Kite, F. E., Maywald, E. C., and Schoch, T. J.,** Functional Properties for starch in foods, *Starke,* 15, 131, 1963.
51. **Sterling, C.,** Molecular association of starch at high temperatures, *Starke,* 12, 78, 1960.
52. **Schoch, T. J. and Elder, A. L.,** Use of sugar and other carbohydrate in the food industry, *Adv. Chem. Ser.,* 12, 21, 1955.
53. **Osman, E. M.,** Starch in the Food industry, in *Starch Chemistry and Technology,* Vol. 2, Whistler, R. L. and Paschall, E. F., Eds., Academic Press, New York, 1969, chap. 8.
54. **Knight, R.,** Speciality Food Starches, in *Proc. Int. disciplinary workshop on processing and storage of cassava,* Araullo, A. L., Nestle, B., and Campbell, M., Eds., IDRC, Ottawa, 1974.
55. **Compton, J. and Martin, W. H.,** Starch in the textile industry, in *Starch Chemistry and Technology,* Vol. 2, Whistler, R. L. and Paschall, E. F., Ed., Academic Press, New York, 1967, 7.
56. **Seydel, P. V.,** *Warp Sizing,* W.R.C., Smith Pub. Co., Atlanta, Ga., 1958.
57. **Dickerson, B. W.,** Use of C.M.C. for textile sizing, *Ind. Wastes,* 1, 10, 1955.
58. **Ananthan, T. V.,** Application of tapioca starches in textile sizing, *Starch and Sago Seminar,* Rotary Club of salem, Salem, India, 1982.
59. **Moorthy, S. N.,** Viscosity and rheological properties of cassava, maize blends, Proc. Natl. Symp. on Root and Tuber Crops, Trivandrum, India, 1986.
60. **Moorthy, S. N.,** Nature and properties of cassava starch subjected to steam pressure treatment, Proc. of Seminar on Post Harvest Technology of Cassava, AFST (I), Trivandrum, India, 1980, 68.

61. **Abraham, E. T., Raja, K. C. M., Sreemulanathan, H., and Mathew, A. G.,** Improvement of cassava flour by chemical treatment, *J. Root Crops,* 5, 11, 1979.
62. **Srivatsava, H. C. and Patel, M. M.,** Viscosity stabilisation of tapioca starch, *Starke,* 25, 17, 1973.
63. **Rutenberg, M. W. and Solarek, D.,** Starch derivatives: production and uses, in *Starch Chemistry and Technology,* 2nd ed., Whistler, R. L., Bemiller, J. N., and Paschell, E. F., Ed., Academic Press, New York, 323, 1984.
64. **Harsveldt, A.,** Starch requirement for paper coating, *Tappi,* 45, 85, 1962.
65. **Kling, A., Traud, W., Werner, H., Mellmut, J.,** Sizing agent for Textiles, *Ger. Offen.,* 2, 426, 1975.
66. **Olsen, H. C.,** Warp sizing of textile yarns with starch amines, *U.S. Patent,* 2, 946, 1960.
67. **Marsh, J. T.,** *Introduction to Textile Finishing,* Chapman and Hall, London, 1948.
68. **Radley, J. A.,** *Starch and its Derivatives,* Vol. 2, Academic Press, New York, 1954, 272.
69. **Mentzer, M. J.,** Starch in the paper industry, in *Starch Chemistry and Technology,* Whistler, R. L., Bemiller, J. N., and Paschall, E. F., Eds., Academic Press, 1984, 18.
70. **Maryanski, J. E.,** Natural binder outlook, presented at TAPPI Annual meeting, February 14-16, CPC International Reports, 1977, 5.
71. **Jones, S. F.,** The world market for starch and starch products with particular reference to cassava (tapioca) starch; G-173, TDRI, London, 1983, 98.
72. **Srivatsava, H. C. and Phadnis, S. P.,** Tapioca starch — problems and potential, Tapioca starch and Sago Seminar, Rotary Club of Salem, Salem, India, 1982, 58.
73. **Houtz, H. H.,** The colloidal differentiation of starches, *Paper Trade J.,* 113, 32, 1941.
74. **Nissen, E. K.,** Starch in the paper Industry, in *Starch Chemistry and Technology,* Vol. 2., Whistler, R. L. and Paschall, E. F., Eds., Academic Press, 1967.
75. **Waters, J. R.,** A technique for evaluation of starch as a wet end additive, *Tappi,* 44, 185A, 1961.
76. **Brill, H. C.,** An evaluation of various beater retention aids for titanium dioxide filler in presence of chlorinated corn starch, *Tappi,* 38, 522, 1955.
77. **Hofreiter, B. T.,** Natural Products for wet end addition, in *Pulp and Paper, Chemistry and Chemical Technology,* Casey, J. P., Ed., John Wiley & Sons, New York, 1984, 1975.
78. **Tessler, M. M.,** Starch ether derivatives, U.S. Patent 3880832, 1975.
79. **Mehltretter, C. L.,** Production and use of dialdehyde starch, in *Starch Chemistry and Technology,* Vol. 2, Whistler, R. L. and Paschall, E. J., Eds., Academic Press, New York, 1967, 18.
80. **Hammerstrand, G. E., Hofreiter, B. T., and Mehltretter, C. L.,** Determination of the extent of reaction between epichlorohydrin and starch, *Cereal Chem.,* 37, 319, 1960.
81. **Harsveldt, A.,** Starch requirement for paper coating, *Chem. Ind.,* 2062, 1961.
82. **Chusing, M. L.,** *Paper and Paper Chemistry and Technology,* Casey, J. P., Ed., John Wiley & Sons, New York, 1981, 1667.
83. **Zuderveld, H.,** in *Chemistry and Industry of Starch,* Radley, J. A., Ed., Applied Science Publishers, London, 1976, 107.
84. **Beals, C. T.,** *Dry Strength Adhesives,* Reynolds, W. F., TAPPI Press, Atlanta, Ga., 1980, 33.
85. Code of Regulations, Title 21, Part 178, Indirect Food Additives, Section 178, 3520, Industrial starch — Modified, U.S. Government Printing Office, Washington, D.C., 1981, 283.
86. **Kline, J. E.,** An investigation of adhesive-pigment interaction in coating mixture, *Tappi,* 55, 556, 1972.
87. **Brown, G. H. and Mazzarella, E. D.,** Sizing of paper, *Ger. Offen.,* 19, 415, 1969.
88. **Lucas, D. E. and McDowell, C. J.,** Addition of corn starch at wet end requires study of many factors, *Paper Trade J.,* 141, 22, 1957.
89. **Lucas, D. E. and Fletcher, C. H.,** New Oxidised starches for paper industry, *Paper Ind.,* 40, 810, 1959.
90. **Srivatsava, H. C. and Patel, M. M.,** Oxidised starch for paper industry, *Tappi,* 54, , 1971.
91. **Buttrick, G. W. and Eldred, N. R.,** Paper containing sulphonated 2-phenoxy ethyl acrelate polymer, *Tappi,* 45, 1891, 1962.
92. **Hershey, R. V. and Hein, G. M.,** Porous smooth coated paper using high solids based coating composition in blade coating operations, U.S. Patent 4,154,899, 1979.
93. **Benninga, M., Harsveldt, A., and De Sturler, A. A.,** Improvement of the stability of clay-satin white coating colours with starch phosphate ester, *Tappi,* 50, 577, 1967.
94. **Mazzarella, E. D. and Hickey, L. J.,** Development of cationic starches as paper coating binders, *Tappi,* 49, 526, 1966.
95. **Powers, R. M. and Bert, R. W.,** Carboxyl starch amine esters as binders for paper coating operations, U.S. Patent 3,598,623, 1971.
96. **Cescato, R. W.,** Starch derivatives for making paper coating compositions, U.S. Patent 3,654,263, 1972.
97. **Dux, E. F. W.,** Production and uses of starch adhesives, in *Starch Chemistry and Technology,* Vol. 2, Whistler, R. L. and Paschall, E. F., Eds., Academic Press, New York, 1967, 23.
98. **Courtonne, H.,** The opposite affect of chlorides and sulfates on starches, *Compt. Rend,* 171, 1168, 1920.

99. **Radley, J. A.**, Adhesives from starch and dextrin, in *Starch and its Derivatives*, Vol. 2, John Wiley & Sons, New York, 1954, 9.

100. **Dulac, R. and Rosenbaum, J. L.**, *Industrial Cold Adhesives*, Griffin and Co., London, 1937.

101. **Bovier, E. M. and Carter, J. A.**, Starch adhesive compositions containing an oxidised waxy starch ester, U.S. Patent 4,231,803, 1980.

102. **Hoepke, C. H., Muller, A., and Ruge, H. J.**, Adhesives for corrugated paper board, Ger. Patent, 2,345,350, 1975.

103. **Nakazawa, H., Shinitini, T., and Okamoto, S.**, Neoprene latex adhesives having good initial adhesive strength, Jap. Patent, 74,648, 1977.

104. **Evans, R. B. and Wurzburg, O. B.**, Production and use of starch dextrins, in *Starch Chemistry and Technology*, Vol. 2, Whistler, R. L. and Paschall, E. F., Eds., Academic Press, New York, 1967, 11.

105. **Kennedy, H.M. and Fischer, A. C.**, Starch and dextrins in prepared adhesives, in *Starch Chemistry and Technology*, 2nd Ed., Whistler, R. L., Bemiller, J. N., and Paschall, E. F., Eds., Academic Press, New York, 1984, 20.

106. **Thompson, A. and Wolfrom, M. L.**, *J. Am. Chem. Soc.*, 80, 6618, 1958.

107. **Brimhall, B.**, Structure of Dyrodextrins, *Ind. Eng. Chem.*, 36, 72, 1944.

108. **Radley, J. A.**, Dextrin and British gums, in *Starch and its Derivatives*, Vol. 2, John Wiley & Sons, New York, 1954, 6.

109. **Fetzer, W. R., Crosby, E. K., and Fullick, R. E.**, Viscometer for dextrins, *Anal. Chem.*, 24, 1671, 1952.

110. **Kerr, R. W. and Schink, N. F.**, Pyrodextrins in coating operation, *Paper Trade J.*, 120, 145, 1945.

111. **Johnson, L. E., Koepsell, H. J., and Churchill, B. W.**, U.S. Patent 2,957,810, 1958.

112. **Smith, P. S.**, Starch derivatives and their use in foods, in *Food Carbohydrates*, Linebeck, D. R. and Inglett, G. E., Eds., AVI Publishing, Westport, Conn., 1982, 14.

113. **Albrecht, J. J., Nelson, A. I., and Steinberg, M. P.**, Characters of cornstarch and starch derivatives as affected by freezing, storage, and thawing, *Food. Technol.*, 14, 57, 1960.

114. **Kerr, R. W. and Cleveland, F. C.**, U.S. Patent 2,884,413, 1959.

115. **Robert, H. J.**, Starch derivatives, in *Starch Chemistry and Technology*, Vol. 2, Whistler, R. L. and Paschall, E. F., Eds., Academic Press, New York, 1967, 13.

116. **D'Ercole, A. D.**, U.S. Patent 3,669,687, 1972.

117. **Antinori, J. A. and Rutenberg, M. W.**, U.S. Patent 3,208,851, 1965.

118. **Radley, J. A.**, Glucose and maltose, in *Starch and its Derivatives*, Vol. 2, John Wiley & Sons, New York, 1954, 4.

119. **Lloyd, N. E. and Nelson, W. J.**, Glucose and fructose containing sweeteners from starch, in *Starch Chemistry and Technology*, 2nd ed., Whistler, R. L., Bemiller, J. N., and Paschall, E. F., Eds., Academic Press, New York, 1984, 21.

120. **Kerr, R. W.**, *Chemistry and Industry of Starch*, 2nd ed., Kerr, R. W., Ed., Academic Press, New York, 1950, 14.

121. **MacAllister, R. V.**, Nutritive sweeteners made from starch, *Adv. Carbohydr. Chem. Biochem.*, 36, 15, 1979.

122. **Fukushi, T. and Isemura, T.**, Regeneration of the native three dimensional structure of *Bacillus subtillis* — amylase and its fermentation in biological systems, *J. Biochem.*, 64, 283, 1968.

123. **Barber, S. A., Somers, P. J., Epton, R., and McLaren, J. V.**, Crosslinked polyacrylamide derivatives as water insoluble carriers for amylolytic enzymes, *Carbohydr. Res.*, 14, 287, 1970.

124. **Marshall, R. O. and Kooi, E. R.**, Enzyme conversion of D-glucose to D-fructose, *Science*, 125, 648, 1957.

125. **Antrim, R. L., Colilla, W., and Schnyder, B. J.**, *Applied Biochemistry and Bioengineering*, Vol. 12, Wingard, L. B., Katchalski-Katzir, E., and Goldstein, L., Eds., Academic Press, New York, 1979, 97.

126. **MacAllister, R. V.**, in *Immobilised Enzymes for Food Processing*, Pitcher, W. H., Jr., Ed., CRC Press, Boca Raton, Fla., 1980, 81.

127. **Spenser, H. W.**, *Sweetness and Sweeteners*, Birch, G. G., Green, L. F., and Coulson, C. B., Eds., Applied Science Publishers, London, 1971, 112.

128. **Kooi, E. R. and Armbruster, F. C.**, Production and use of dextrose, in *Starch Chemistry and Technology*, Vol. 2, Whistler, R. L. and Paschall, E. F., Eds., Academic Press, New York, 1967, 24.

129. **Radley, J. A.**, Miscellaneous uses, in *Starch and its Derivatives*, Vol. 2, Radley, J. A., Ed., John Wiley & and Sons, New York, 1954, 13.

130. **Kolaian, J. H. and Park, J. H.**, Oxidised starch as a dispersant, U.S. Patent 3,417,017, 1968.

131. **James, R. W.**, *Industrial Starches*, Noyes Data Corporation, New Jersey, 1974, 304.

132. **Burbank, H. H., Brake, J. M., and Roberts, M.**, Plasma expanders, U.S. Patent 3,523,938, 1970.

133. **McKillip, W. J., Kellen, J. N., Ipola, C. N., Buckney, R. W. and Otey, F. H.**, Coatings from glycol glycosides in alkyds, *J. Paint. Technol.*, 42, 312, 1970.

134. **Otey, F. H. and Doane, W. M.,** Chemicals from starch, in, *Starch Chemistry and Technology,* Whistler, R. L., Bemiller, J. N., and Paschall, E. F., Eds., Academic Press, New York, 11.

135. **Otey, F. H., Westhoff, R. P., Kwolek, W. F., Mehltretter, C. L., and Rist, C. E.,** Urethane plastic based on starch and starch derived glycosides, *Ind. Eng. Chem. Prod. Res. Dev.,* 8, 267, 1969.

136. **Weaver, M. O., Fanta, G. F., and Doane, W. M.,** Proc. Tech. Symp. Nonwoven Product Technol., Intern. Nonwovens disposables Assoc., Washington, D.C., 1974, 169.

137. **Smith, T.,** Starch acrylonitrile graft copolymer, U.S. Patent 3,661,815, 1972.

Chapter 10

CASSAVA BASED INDUSTRIES

Cassava is one of the richest sources of starch. The tuberous roots contain up to 30% of starch and are low in proteins, soluble carbohydrates, and fats. Extraction of starch from cassava, consequently, is a simple and straightforward process without the problems associated with the manufacture of corn, wheat, or other cereal starches. The process is unique in the sense that it is applied at cottage, small, and large scale levels of production. The cassava starch industry is important enough to warrant the attention of both researchers as well as exporters in developing countries, considering the fact that about 85% of the export of starches by these countries is accounted for by cassava starch, though the share of cassava starch in world starch production is only 8%.[1,2]

I. CASSAVA STARCH PRODUCTION

The roots of cassava should be processed within 24 hr after harvesting. The most essential factor in the production of good grade cassava starch is that the whole process, from harvesting the roots to completion of the final drying, should be carried out in the shortest time possible, since deterioration sets in from the time of root extraction and proceeds throughout the process.

Basically, cassava starch manufacturing can be divided into the following stages:

1. Washing and peeling of the tubers to remove and separate all adhering soil and as much protective epidermis as necessary.
2. Rasping or disintegration to destroy the cellular structure and to rupture the cell walls to release the starch as discrete, undamaged granules from other insoluble matter.
3. Screening or extraction to separate comminuted pulp into two fractions, viz. waste fibrous material and starch milk.
4. Purification or dewatering to separate the solid starch granules from their suspension in water by sedimentation or centrifuging.
5. Drying to remove sufficient moisture from the damp starch cake obtained during the separation stage so as to reduce the moisture content from 14 to 35% to 12 to 14%, a level low enough for long-term storage.
6. Finishing operations such as pulverizing, sifting, and bagging.

The manufacture of cassava starch is, as stated earlier carried out in roughly three types of establishments. The first is the cottage industry, where the work is carried out entirely by rudimentary hand tools, usually operated by a single family and producing 50 to 60 kg of crude starch per member per day. The second type, small scale enterprise, produces about 5 to 40 t of tubers per day, mainly because of more efficient rasping obtained by use of a prime mover of about 20 hp and needing little skilled labor. The third type is the large scale factory which may sometimes operate its own extensive plantations, thus assuring a regular supply of raw materials, processed using modern equipment. The third type of mill processes about 100 t of tubers or more per day.[3-5]

A. Cottage Scale: Manual Methods of Starch Extraction

The cassava roots are washed by hand and peeled with handknives. These are then manually rasped to a pulp on a stationary grater which is simply a tin or mild steel plate perforated by nails so as to leave projecting burrs on one side. The pulp is taken on a piece of fabric fastened on four poles and washed vigorously with water by hand. Finally the fiber is squeezed out while the starch milk collects in a bucket. When starch granules settle down the supernatant water is decanted and the moist starch is crumbled and dried in a tray or on bamboo mat. In some places the starch milk is squeezed through a closely woven thick fabric to trap the starch granules or hung overnight to remove gravitational water which is then sun-dried. This simple process is used by many natives in rural areas of the tropical regions.

B. Small Scale Industrial Starch Manufacture: Semimechanized

1. Preparation

In small-scale operations and where labor is cheap and plentiful, washing and peeling can be very adequately accomplished by hand. Usually peeling is carried out with the help of ordinary knives. However, as work must be finished as quickly as possible, numerous hands need to be employed. For production of edible starch, the outer peel (skin) as well as the sub-periderm (rind) are removed (which would otherwise entail difficulties in rasping and also in removing dirt, crude fiber and cork particles). The sub-periderm, usually 2 to 3 mm thick, contains only 50% of the starch as the core of the tuber does and it also contains most of the hydrocyanic acid (HCN), which may cause some discoloring of the starch.[6] However, in larger factories, only the outer skin or corky layer is removed as it is profitable to recover the starch from the cortex or rind, which represents about 8 to 15% of the whole root by weight.

At this level, the roots are usually only washed after peeling. A marked improvement would be possible for the roots were washed before and after peeling. In fact, experiments have shown that washing-peeling-washing can cut ash content of the finished product in half.[7]

2. Rasping

Simple but effective root disintegration is obtained by a rasper. It is a wooden drum with a steel shaft and cast iron ends. A sheet of metal, perforated with nail, is clamped around the drum with the protusions facing outside. The drum rotates in a housing with a hopper at the top for feeding the peeled and washed tubers and with a perforated metallic plate underneath, through which the sufficiently rasped pulp has to pass into the sump below. A small quantity of water is continuously added in the rasper. Power for turning the rasper and also for actuating the shaking screens is usually supplied by an electric motor through countershaft and a necessary belt drive arrangement.

The perforated rasping plate, though inexpensive, is relatively inefficient as the rasping plate must often be replaced on account of rapid wearout. It is difficult to remove all the starch, even with efficient rasping devices, in a single operation. Therefore, the pulp is sometimes subjected to a second rasping.

a. Rasping Effect

Against the sharp protusions of the rasper surface, the cell walls are torn up and the whole of the root flesh is turned into a fine pulp in which most, but not all of the starch granules, are released. The percentage of starch set free by rasping is called the rasping effect (R) which may be 70 to 90% after one rasping.

The rasping effect (R) may be directly evaluated by washing out a weighed sample of the pulp obtained from the factory rasper on a 260-mesh sieve, collecting the starch on a filter and weighing it after drying, and then dividing this fraction by the total starch content of the pulp.

To obviate the rather difficult direct determination of free starch, the following method may be preferred by using only the analysis for starch and fiber contents of the roots being processed and of the waste pulp produced in processing them.[4] Since the waste pulp, as

FIGURE 1. Final washing of screened starch milk.

also the root, consists mainly of water, cellulose, and starch, the percentage of starch can be deduced from the moisture and the fiber contents, further simplifying the method.

If the starch contents of the roots and the waste pulp are S_r and S_w, respectively, and the corresponding fiber contents are F_r and F_w, it is obvious that the fraction of the starch remains bound to the fiber by escaping from rasping amounts to $(S_w/F_w)/(S_r/F_r)$. Thus,

$$R = [1 - (S_W \cdot F_R)/(S_R \cdot F_W)] \times 100 \%$$

High rasping effects involve a lower use of energy while the influence of peeling the root is favorable for both the aspects.[4]

3. Screening

After rasping, pulp from sump is pumped on to a series of flat, slightly inclined vibrating screens of diminishing mesh size. The screens used are usually three in number of 80, 150, and 260 mesh, respectively,[3,4] with the first retaining the coarse fiber and the other fine particles. Usually a small spray of water is applied to assist the separation of starch granules from their fibrous matrix and to keep the screen meshes clean. Starch granules carried with water fall to the bottom of the tank in which the sieves are placed from where the starch milk is channelled for gravitational sedimentation. Sometimes a final washing is carried out manually over a 300 mesh screen (Figure 1). Nylon, phosphor bronze or stainless steel wire mesh are used for screens.

In separating the pulp from the free starch a liberal amount of water must be added to the pulp and then stirred vigorously. In "wet screening" both the operations are combined.

Residual pulp from the screening is considered as a by-product of the cassava starch industry. It is about 10% by weight of the cassava roots, and on the basis of dry matter consists of about 56.0% starch, 35.9% fiber, 5.3% protein, 2.7% ash, and 0.1% fat. It is used as an animal feed, either wet (75 to 80% moisture content) in the neighborhood of the factory or sun-dried before sale.

4. Dewatering
a. Settling Tank

The oldest practice for settling starch from its suspension in water is to let the starch milk

FIGURE 2.   Settling table for sedimentation of starch.

stand for a period of 8 hr in tanks with plugged effluent outlets at varying heights. Operated batchwise, when the starch settles down at the bottom of the tanks, the supernatant liquor is run off. During this process of dewatering, a number of tanks are usually filled in succession. The dimensions and the number of tanks are determined by the level of production and the convenience of handling.

The upper layer of the sedimented starch cake which has a yellowish green tint, contains many impurities and is usually scraped off and rejected. In most cases, two settlings suffice to obtain a reasonably clean flour. In India, however, even the last traces of starch in the supernatant water, accumulated in a final settling basin, are collected after a couple of days which is known as *azhukkumavu*, meaning grey or dirty starch.

b. Settling Table

Gravitational sedimentation can also be achieved in a semicontinuous process on settling tables. In the better equipped factories the simple settling tank is giving way to the settling tables. The table consists of successive sets of channels or troughs, about 40 to 50 cm wide, 20 to 30 cm deep, and 30 to 100 m long. The starch suspension is allowed to flow very slowly along slightly inclined shallow troughs. When sufficient starch settles out at the base of the channel, the flow of starch milk is temporarily stopped and the starch removed manually (Figure 2).

In settling tables there is considerable reduction in time of contact between the fruit water (water containing soluble carbohydrates and nitrogenous compounds, sometimes also referred to as ''latex'') and the starch. The starch settles on different parts of the table differentiated according to purity and granule size, with the higher end containing more of the larger starch granules and very little of protein and other contaminants. Thus it enables the manufacturer to simultaneously produce at least two grades of starch without any extra cost.

c. Silting

Silting is the term applied to the separation of starch from starch milk on more steeply inclined tables than those used in the tabling process. In the tabling process, the suspension generally contains 25 to 30 g/ℓ, but in silting higher concentrations up to 250 g/ℓ can be allowed.[3]

d. Effect of Fruit Water

The quality of starch produced depends to a great extent on the proper performance of the whole series of operations for separating the pure starch from soluble contaminants. They result in a more or less concentrated suspension of starch in pure water. As has been stated earlier, the entire processing of cassava must be completed within as short a time as possible. This is particularly true for the separation of free starch from its suspension in the so-called fruit water, because of the very rapid chemical changes in this solution and formation of very stable complexes between starch and proteins, fatty materials, etc., from which it is almost impossible to separate the pure starch.

At a later stage, the fruit water being rather rich in sugar and other nutrients, microorganisms start to develop and eventually lead to a vigorous fermentation. Alcohols and organic acids are formed, among which butyric acid is particularly noticeable because of its odor.[4] Indeed, small rural mills can often be located by the smell of the butyric acid. These biochemical changes exert a negative influence on the quality of the flour similar to the preceeding physico-chemical changes.

e. Effect of Granule Size on Settling

A spherical particle suspended in a fluid medium soon acquires, under the influence of gravity, a motion of constant velocity (known as the terminal velocity) which, based on Stoke's law,[8] is given by

$$u_t = \frac{g\, D_p^2\, (f_p - f)}{18\, \eta \times 10^8}$$

where u_t = velocity of the particle, cm/s; D_p = diameter of the particle; f_p = density of the particle, g/cc; f = density of the medium, g/cc; η = viscosity of the medium, Poise; and g = acceleration due to gravity; cm/s².

In cassava starch milk, which consists of spherical particles of approximately 1.5 sp. gr. in a medium with a viscosity not much different than that of water, the speed of sedimentation will primarily depend on the diameter of the granules.

The size of the cassava starch granules is not exactly uniform, the diameter ranging between 4 and 24 μm. In the successively deposited layers of sediment, therefore, a gradation occurs according to the granule size as larger particles attain higher terminal velocity. However, the gradation intensifies with the length of the path of sedimentation, and lower layers of the tank contain granules of a wide range of sizes, settled during the first stages.[4]

In settling tables, similar gradation is to be expected, but with respect to the distance from the head of the table. Centrifuged separators in modern installations, however, produce a uniform mixture of starch granules of all sizes which occur in the starch milk.

f. Effect of Chemicals

In sedimentation, the compactness and the consistency of the settled starch layer is of major importance. The starch losses incurred with draining of the supernatant, water will decrease as the starch settles to a firm cake. Pure starch settles in clean water to a compact mass showing good dilatency,[3-4] i.e., it crumbles when broken but thereafter it loses all form and spreads like a thick syrup as the deforming force is withdrawn, a phenomenon observed while scooping the wet starch cake. This occurs since starch granules in the sediment, when at rest, naturally pile up in the most space-saving manner, whereas any disturbance of this array by external forces results in an increase of the interstitial volume.

The compactness and also the volume of the sediment depend very much on the presence of impurities, such as fiber, which tends to result in a softer sediment. Some chemical aids can be used to improve the consistency of the sediments. An acid reaction of the suspension promotes rapid settling and compact sediment, while an alkaline reaction has the opposite effect. However, it should be emphasized that by providing the basic fast and clean working conditions, there should be no need to use the chemical additives.

Sulfuric acid, added as an aid to sedimentation, produces a whiter starch. Its effect on sedimentation is noticeable, at concentrations above 0.001 mℓ of the concentrated acid (specific gravity 1.84) per liter of starch milk of 2° Brix (equivalent to about 2 g starch per liter of water). Addition of 0.01 mℓ acid gives rapid sedimentation, but rather a soft sediment and a strong decrease in the viscosity of the starch paste. Up to about 0.001 mℓ/ℓ, dose, however there appears to be a slight increase in viscosity.

Sulfur dioxide is usually added to most grain starches, but its advantage in the processing of root starches is debatable.[4] It helps to check bacterial and enzymatic action but it can reduce the viscosity of the product. Sulfur dioxide also acts as a bleaching agent, although the white color thus obtained soon deteriorates.

Alum or aluminum sulfate may be present in the starch milk due to over dosage during purification of the process water. It has a favorable effect on sedimentation. An addition of 0.1 g/ℓ to starch milk of 2° Brix is said to result in an increase of approximately 50% in viscosity.[3]

Chlorine, in its elemental, chloride, or hypochlorite form, considerably augments the viscosity of the product, provided the concentration is kept below 1 mg/ℓ starch milk.[4] At this level, the sediment obtained is very compact and white, but higher concentrations of approximately 50 mg/ℓ result in a very soft and discolored sediment in addition to lowering the viscosity of the product.

5. Drying

After the removal of free water from the starch by sedimentation, a cake is obtained containing 35 to 40% moisture. Usually the starch cake is crumbled into small lumps (1 to 3 cm) and spread out in thin layers (less than 3 cm) on large open areas for sun-drying. It is either deposited on concrete yards or on wooden trays raised to about 1 m above the ground to allow air circulation. Drying generally takes 24 to 120 hr[7] during which time dirt contamination is a real problem resulting in occurrence of "specks" and lowered whiteness. However, an important advantage of the sun-drying is the bleaching action of the ultraviolet rays. The crude starch is considered sufficiently dry when the lumps are too hard to be crumbled by hand and has a moisture content between 15 and 20%.

In Malaysia and other parts of the Far East, ovens called "drying yards" are used for drying cassava starch.[4] The cemented drying floors are about 30 to 40 m long and 3 to 5 m wide and consist of a firing tunnel of brickwork covered with galvanized iron or copper plates, where enough wood is burnt to generate heat. Firing has to be moderate to keep the surface temperature well below the gelatinization point of the starch.

Tray dryers, known as chamber dryers, are also in use for drying cassava starch in some small- and medium-size factories producing limited quantity of starch. Improvement in quality of starch is reported with mechanized drying (50°C, 6 hr) which lowers ash content from 0.20 to 0.14%.[7] A completely satisfactory solution to the starch drying problem for medium factories is, however, still elusive.[3]

After drying the starch, the agglomerates are powdered with the help of roller grinder or pulverizing mill followed by dry-screening. The latter operations are often referred to as bolting. The revolving screens used at this stage to sift the foreign particles and ensure lump-free uniform product have a screening gauze of 100 to 200 mesh. The finished starch is then bagged. Packaging and storage aspects of cassava starch have been already discussed in detail in a previous chapter (Chapter 5).

The major technological problems associated with the cassava starch industry in India, and elsewhere, are high water consumption, large space requirements, long detention time, and weather-dependent drying adversely affecting the product quality.

C. Modern Methods of Cassava Starch Extraction

Roots in the larger factories, on being received, are immediately peeled and washed by mechanical scrubbing. The washing machine is a perforated drum partially immersed in a water bath. The roots are propelled forward by a series of paddle arms, or a spiral brush, attached to a central rotating shaft. A counter-current flow of water through the bath ensures continuous removal of dirt. In some designs high pressure water sprays from nozzles may also act on the roots. Peeling is accomplished often as an extension of the washing machine. The combined action of the high pressure water jets and abrasion of the tubers against the drum walls and against each other remove most of the skin.[9]

A jahn-type rasper, used in the modern process, consists of a rotating drum of about 40 to 50 cm diameter and 30 to 50 cm length with longitudinally arranged sawtooth blades in grooves milled around the circumference. Blades have 8 to 10 sawteeth per cm and are spaced 6 to 10 mm apart projecting about 1 mm above the surface. The optimum speed is 1000 rpm,[3] corresponding to a linear velocity of about 25 m/sec. In many mills, the coarse pulp retained on first shaking screen is reground in a secondary rasp with finer blades having a greater number of teeth per unit length of blade (10 to 12 per cm), and then returned for rescreening. While a rasping effect of about 85% is achieved at the first rasping, the overall rasping effect is raised to 90% after secondary rasping.

Shaking screens can be used on a larger scale, in a series of increasing fineness, such as 80, 150, and 260 mesh phosphorus bronze, respectively, aided by gentle wash water sprays. However, modern practice is to use the sieve bends or DSM screens, working in 3 to 6 stages in a series. The development of sieve bends is recognized as one of the biggest in starch manufacture. The sieve bend is a stationary curved screen constructed of wedge section stainless steel wires or rods with average slot width of 75 μm between them.[6] The preslurried pulp is sprayed at a right angle to the wedges. Flowing down across the screen, the smaller starch granules pass through the slots and the larger fibrous material are separated continuously. A countercurrent system of overflow does extraction and sieving, requiring no fresh water for washing. Rotating screens, usually horizontal sieve cones, by the action of centrifugal force make the retained fiber slide over the screen and fall out. Rotating screens, also used sometimes in large scale operations, may operate batchwise or continuously.[9]

After being separated from fibers, the starch needs to be dewatered. Mechanical dewatering is commonly done either on vacuum filters or centrifuges. Vacuum filter consists of a slowly rotating perforated hallow horizontal cylinder and is covered by a cloth. The cylinder is partially submerged in a trough containing starch suspension. A vacuum pull within the cylinder sucks the water while starch adheres to the cloth screen which is scraped out continuously.

Centrifuges of numerous designs are used in the starch industry. The most common type of the sedimentation centrifuges is the disk centrifuge. Feed is admitted to the center of the bowl near its floor and rises through a stack of sheet metal "disks", which are actually truncated cones spaced 0.4 to 3 mm apart.[7] The half angle made by the disks along with the vertical is typically between 35 and 50°. Each disk carries several holes, 6 to 13 mm in diameter when assembled in place in the bowl; these disks form several channels through which the liquid rises.

The purpose of the disks is primarily to reduce the sedimentation distance by splitting up the liquid in many layers. Once the solid particle reaches the underside of one of the disks, it is in effect removed from the liquid. It continues to move outward until it is deposited on the peripheral wall of the bowl. Clarified effluent may be discharged from the bowl through overflow ports. Accumulated solids may be removed periodically by hand after stopping the machine and dismantling it.

Nozzle type separators, possibly the most important development of the basic design of disk centrifuge used in the starch industry,[3] enable collection of the solids by continuous discharge. The periphery of the bowl is fitted with nozzles, varying between 6 and 12 in number and between 0.8 and 2 mm in size. Dilute starch slurries may be concentrated up to rates of 70 m³/hr to obtain concentrations up to 270 g/ℓ of starch,[3] and losses in the effluent of as little as 1 to 2 g/ℓ.

Peeler centrifuges operate on a horizontal axis in which a hydraulically operated knife peels off the starch separated after drainage of the mother liquor and wash water. These centrifuges are used for dewatering during cassava starch extraction. In the case of over-dilution during screening, the starch suspension may be preconcentrated by gravitational sedimentation or by hydrocyclones and thus reduce the volume and size of final equipment. Mechanical separators are fast and compact, slashing off residence time and floor area required.

To speed the drying process, independently of weather conditions, tray dryers, rotary driers, and belt-and-tunnel dryers have been used followed by grinding of the dried product. However, for the largest scale operations, flash or pneumatic dryers are used. The damp starch is transported, in a vertical stream of hot air at a temperature of about 150°C, to a cyclone filter, where dried starch granules are separated from air. Residence time is a few seconds and the starch from a flash dryer is now a fine powder, with a final moisture content of 10 to 13%.

1. Alfa-Laval Process

In the most modern processes, particularly the Alfa-Laval system, the process time from root to finished starch is only 40 min.[3] In this process, after the usual preliminary steps, the pulp is passed to a Rotasieve (high-speed rotating vertical cone sieve) from which the starch is washed as a milk of 2 to 3° Be'. After secondary rasping, the pulp is resieved in a second Rotasieve and the exhausted material discharged to waste. The combined starch milks are fed to a nozzle-type centrifuge in which some fine fiber is removed; the starch is washed and concentrated to 11 to 12° Be'. After dosing with 0.2% sulfuric acid at the rate of about 6 mℓ/kg of starch solids in suspension and sieving, the milk passes to another nozzle centrifuge for further washing and is concentrated to 19° Be'. From this centrifuge, finally it passes to a purification centrifuge in which the impurities are skimmed off and the starch gets washed, reslurried, and discharged at 18 to 20° Be' to a holding tank. From the holding tank it is either fed to a vacuum filter or to a dewatering centrifuge. The moist starch cake from either of these machines is then dried in a flash dryer.

In this process, by careful reuse of the wash waters, process water requirement can be reduced to 32 m³/t of starch.[10] The final effluent may be returned to the soil via irrigation channels to replace some of the nutrients.

2. Dorr-Oliver Process

In yet another modern process, the Dorr-Oliver system, after peeling, washing, and disintegration of the tubers, the fruitwater is removed in solid bowl centrifuges.[6] Removal of the fruitwater right at the start of the process eliminates chances of fouling the starch. The fibers are then removed on a small battery of DSM screens in usually 3 or 4 stages depending on economy. The fibrous overflow of the final screening stage has a dry solids (DS) content of about 35 to 40 g/ℓ. Depending on local circumstances this refuse may be dewatered in plate and frame presses or disposed of as such for animal feed or fertilizer. After screening, the crude starch milk is degritted, concentrated, and purified in a counter-current Dorrelone washing system with a small amount of wash water (3 kg of fresh water per kg of DS starch). The absence of any moving parts in this system, except from the pumps, considerably reduces the maintenance costs. The starch is next dewatered on a rotary vacuum "heel cake" filter with scraper discharge and dried in a flash dryer. Based on this system, the crude water consumption is about 2000 m³/day, of which at least 75% can be reused, and the softened water consumption is about 700 to 750 m³/day for a cassava starch plant grinding 250 to 300 t/day of tubers.[6]

D. Extraction of Starch from Dried Cassava

Rapid deterioration of fresh tubers hindering year-round supply, and also the fact that there is a substantial reduction in the HCN content, has necessitated attempts for the preparation of starch from the dried root. However, the process is not only difficult, but also the viscosity characteristics of the starch so prepared is inferior to those of starch from fresh roots.[3]

Cassava roots are coarsely subdivided and converted to a crushed, long-fibered mass by repeated pressure treatment between rollers that rotate with equal peripheral speed so that little powder is produced. The mass is then soaked in water and further treated to extract the starch.

A limited quantity of the cassava imported into Europe in the form of chips and dried, slice roots is manufactured into starch.[4] The dried roots are cleaned, washed, and grated, and the starch is separated by cylindrical sieves. However, this practice is expensive and the starch produced is of inferior quality. The brown skin, which contains chlorophyll and coagulated proteinous substances, adheres strongly to the ligneous tissues, imparting a dark color to the starch of the dried roots. This problem, of course, can be taken care of by using cassava chips prepared from tubers that were peeled prior to drying. The other process difficulty is the nitrogenous substances which are found in a colloidal state enveloping the starch granules. It is easier to separate these particles in the pulp slurry of fresh roots than in dried roots. It is also obvious that the preliminary operations of disintegration-dehydration-disintegration make the process highly energy consuming and hence expensive.

E. Physical Methods of Determination of Starch Content in Fresh Cassava Tubers

Prices of the cassava tubers, in the context of manufacturing starch, are often set on the basis of an optimum starch content (about 25%) with a discount or a premium for deviations from that level. As chemical analysis requires qualified technicians, laboratory facilities, and time, the starch content in the tubers is determined subjectively by the representative of the factory by means of breaking a medium size root in the middle into two. If the tuber snaps with medium force, the crop is generally regarded as mature; with the firm, white, and dry flesh the tuber is considered to have high starch content (about 30%). Immature tubers usually have yellowish and translucent watery flesh with low starch content. In case considerable force is required to snap the tuber, the cassava has become woody and the crop has passed its prime, resulting in poor recovery of starch. Such subjective evaluations can be fairly accurate with experienced personnel. Some physical methods are described below which can prove convenient in handling large number of samples at the field level. Nevertheless, estimation of starch content is best carried out with the use of standard analytical methods.

1. Methods Based on Dry Matter Content

By determining the moisture content of the fresh roots it is possible to estimate the starch content by using the following empirical relations,[4] which can save a great deal of time and exertion.

For whole roots:

$$\% \text{ starch } = \% \text{ dry matter content } - 7.3$$

$$= 92.7 - \% \text{ moisture content}$$

For peeled roots:

$$\% \text{ starch } = \% \text{ dry matter content } - 6.8$$

$$= 93.2 - \% \text{ moisture content}$$

The above relationships were established as the result of a series of analyses of roots of four different strains of cassava, all grown in the same soil and during the same period. Another study made of the relationship between dry matter and starch contents,[12] based on trials evaluating 15 cultivars, gives the following empirical equation valid for peeled roots.

$$\% \text{ starch } = \% \text{ dry matter content } - 7.5$$

It is apparent that the constants appearing in the above equations may have to be determined for each new set of circumstances, viz., varieties, cultural practices, and environments of the crop grown.

2. Methods Based on Specific Gravity

The specific gravity of the roots can be measured by means of a commercially available Reimann balance, sold for use with cassava. The apparatus includes a beam balance from which two wire baskets are suspended over a water-filled container. After weighing a sample of roots in the upper basket, the roots are tipped into the lower basket and immersed in water before weighing a second time. Calibrations on the balance itself provide direct readings of starch content when a predetermined quantity of sample roots are used. In a comparative evaluation, however, calibrations on the Reinmann balance marked by the manufacturers were observed to have grossly underestimated the starch content at the lowest specific gravity values.[13] The Reinmann balance is sufficiently portable for making the measurements in the field while harvesting.

Studies conducted at CIAT, Columbia, have indicated very high correlation between root-specific gravity and root dry matter content, as well as between root-specific gravity and root-starch content. The following regression equations have been shown to represent the linear relationships.[14] For whole roots:

$$y = -1.064 + 1.221x$$

For peeled roots:

$$y = -1.104 + 1.221x$$

$$\text{where} \quad y = \% \text{ starch content}$$

$$\text{and} \quad X = \text{specific gravity of the whole root}$$
$$\text{in both the above cases}$$

These results were based on experiments conducted with three cultivars of cassava.

3. Maceration and Sieving Technique

Krochmal and Kilbride[15] have suggested a simple and inexpensive method for quantitative starch extraction. About 50 g sample of cassava root, fresh or frozen during peak seasons, is sliced and macerated with 500 mℓ of water for 5 min in a kitchen blender. The pulp is

FIGURE 3.   Unpolluted view of the lake.

washed on a 200 mesh sieve with an additional 500 mℓ of water. The washed material, which consists of starch and water, is poured into aluminium pans and dried in a forced-draft drying oven at about 85°C for 6 to 12 hr until a constant weight is attained. The weight of residue as a percentage of the weight of the sample represents the starch content. The method appears to be accurate enough to detect any differences of economic importance to a commercial processor of cassava.

Ramanujam[16] has recommended a similar method for estimation of extractible starch in cassava. However, this method comprises of overnight gravitational sedimentation after the wet grinding and sieving with a 260-mesh screen. The final drying is carried out after decanting the supernatant water. The method is energy saving in comparison to that described by Krochmal and Kilbride, but more representative of the industrially recoverable starch rather than the actual starch content of the tuber.

Although chemical methods are the most accurate in starch determination, there are occasions arising in research laboratories as well as cassava starch factories when absolute accuracy of starch estimates can be sacrificed in favor of swiftness to enable the processing of a large number of samples in a short time. Application of these methods may be a great help under such circumstances.

II. EFFLUENTS AND WASTES FROM CASSAVA STARCH FACTORIES

The process to extract starch from cassava tubers requires large quantities of water and significant amounts of waste waters (effluents) are released. Most of the starch factories are located near the banks of rivers or lakes and it has become customary to discharge the effluents from the factory to the river or lakes (Figures 3 and 4). These effluents pose a serious threat to the environment and quality of life in the rural areas. The magnitude of such pollution problems in Thailand[17,18] and India[19] have been described by many workers. They have also reported the possibility of utilizing the cassava starch factory effluents for the production of single-cell proteins.

FIGURE 4.   Lake polluted with cassava starch effluents.

Table 1
CHARACTERISTICS OF WASTEWATER SAMPLES FROM A STARCH FACTORY

	Large-scale		Small-scale
Parameters	**Primary concentration range**[a]	**Secondary concentration range**[a]	**Secondary concentration range**[b]
COD	33,600—38,223	3,800—9,050	3,870—6,670
BOD[c]	13,200—14,300	3,600—7,050	3,400—6,018
pH	4.5—4.7	4.5—4.7	4.5—4.7
Free sugar as glucose	425—1,850	735—2,060	640—2,075
Total hydrolyzable reducing sugar as glucose	22,614—29,275	1,120—2,761	1,590—3,019
Total suspended solids	33,200—37,320	980—4,078	1,868—2,960
Total solids	35,640—42,000	3,200—9,600	4,000—6,600
Total ash	1,450—1,680	265—820	310—520
Total nitrogen	97—182	62—86	65—74

Note: All values are in mg/ℓ.

[a]   Data obtained from 24 samples and analysis.
[b]   Data obtained from 2 samples and analysis.
[c]   Assayed after 20 hr.

Wide variations observed in the physical and chemical constituents of primary and secondary effluents obtained from the cassava starch factories.[20] In the case of a large cassava factory in Kerala, India, the chemical oxidation demand (COD) concentration in the primary effluents ranged between 33,600 to 38.223 mg/ℓ, whereas in the secondary effluents, the range was only 3800 to 9050 mg/ℓ (Table 1). Similarly, the level of biological oxidation demand (BOD) also varied from primary to secondary effluents. In the primary effluents

Table 2
TOTAL MICROBIAL FLORA IN THE CASSAVA STARCH FACTORY EFFLUENTS

Samples	Bacteria	Yeasts	Fungi	Actinomycetes
Primary waste (large-scale)	5×10^5	1.2×10^3	2×10^4	3×10^3
Secondary effluents (large-scale)	200×10^5	29×10^3	1.7×10^4	2×10^3
Secondary (small-scale)	10.5×10^5	3.5×10^3	1.5×10^4	2×10^3

Note: Population expressed as $10^x/m\ell$ of sample.

Table 3
MORPHOLOGICAL GROUPING OF BACTERIA, YEASTS, AND FUNGI

Microorganisms	Primary (large-scale)	Secondary (large-scale)	Secondary (small-scale)
Bacteria			
Gram-positive cocci	−	+	−
Gram-positive rods	+	+	+
Gram-negative bacilli	+	+	+
Gram-negative rods	−	+	+
Spore formers	+	+	+
Nonspore formers	+	+	+
Fungi			
Aspergillus sp.	+	+	+
Penicillium sp.	+	+	+
Monocillium sp.	−	+	−
Humicola sp.	+	−	−
Unidentified sp.	+	+	+
Yeasts			
Candida sp.	+	+	+
Geotrichum candida	+	+	+
Unidentified sp.	+	+	+

Note: + indicates present; − indicates absent.

the BOD level was in the range of 13,200 to 14,300 mg/ℓ and the corresponding figures for the secondary effluents were 3600 to 7,050 mg/ℓ. The COD and BOD level of the effluents from small scale starch factory unit varied from 3,870 to 6,670 and 3,400 to 6,018 mg/ℓ, respectively. The increase in COD and BOD levels in the primary effluents were due to the presence of high amount of suspended solids. The COD and BOD level in the starch factory effluents could be reduced by plain sedimentation as reported in the sedimentation studies on separator waste.[3] The acidity of the effluents ranged between 4.5 to 4.7 and the release of hydrocyanic acid (HCN) during the process of extraction of starch from cassava has been reported as the cause for the acidic nature of the effluents.[18] Nitrogen and phosphorus are the main nutrients for the efficient stability of organic wastes, and analysis of the effluents revealed that there was low nitrogen content in both the effluents, which indicated the necessity for enrichment of effluents to reduce the BOD and COD.

Various types of microorganisms are harbored in the effluents, i.e., bacteria, yeasts, fungi, and actinomycetes.[20] An account of various morphological groups of microorganisms are given in Tables 2 and 3. The secondary effluents of large scale starch factory harbored maximum bacterial population (2×10^7) when compared to the primary effluents ($5 \times$

10^5). But the fungal and actinomycetes population were found to be higher in the primary effluents. This could be due to the occurrence of cellulosic materials in the primary effluents which accelerated the growth of cellulolytic fungi and actinomycetes. The yeast population was higher in the secondary effluents (2.9×10^4) when compared to the primary effluents (1.2×10^3). The bacterial population in the effluents of small scale starch factories was also high (10.5×10^3) when compared to the primary effluents of large scale factories.[20] The presence of higher amounts of free sugars in the secondary effluents promoted the growth of yeast and bacterial flora when compared to the primary effluents. The actinomycetes population of the secondary effluents of the small scale factory was at par with the effluents of the large scale starch factory effluents (2×10^3). Coliform bacterial counts are comparatively higher in the effluents (1600/100 mℓ). Compared to gram positive bacteria, the gram-negative organisms were less in all the samples. Both spore formers and nonspore formers were isolated from the effluents of starch factories. Various species of *Aspergillus Penicillium, Monocillium,* and *Humicola* were identified from primary and secondary effluents of large scale starch factory.

Presence of yeast strains like *Candida* and *Trichosporon* in the effluents indicated the possibilities of their utilization for low-cost single-cell proteins.[18,19,21]

Waste residues amount to approximately 20% of the weight of roots processed is produced by the small and large scale industries manufacturing starch and sago from cassava roots or chips in India. Usually the waste residue is sun-dried and sold as cattle feed. The waste residue containing 60 to 70% starch on a dry weight basis have been successfully saccharified by acid-enzyme or enzyme-enzyme hydrolysis to obtain glucose and is utilized as a substrate for the alcoholic fermentation.[21] The studies revealed about 98 to 99% conversion of dry starch to reducing sugars by enzyme-enzyme saccharification. Fermentation of the saccharified waste after fortification with nitrogen and minerals resulted in a 30 to 33% alcohol yield based on sun-dried waste residue.

III. LIQUID GLUCOSE AND DEXTROSE

Starch is a polymer of glucose and hence is the raw material for glucose. The hydrolysis of starch to glucose can be carried out by acid hydrolysis or enzyme hydrolysis. In India mostly acid hydrolysis is carried out, whereas in other countries enzymes are being used in increasing amounts for production of glucose syrup. The syrup obtained may be further purified and dextrose crystals can be crystalled out. However, glucose syrup itself has a wide range of applications and thus is always in high demand.

The steps for the production of glucose syrup from cassava starch is given in Figure 5.

Starch is suspended in water, of approx. 25 to 30%, solids, to which sufficient HCl has been added to bring a normality of 0.01 to 0.02 HCl. It is heated in a converter under a pressure of 30 lb/in.^{2} for 15 min (140 to 160°C). The reaction mixture is periodically tested for residual starch by iodine staining. When no color develops with iodine, the heating is stopped, pressure is released, and liquid is transferred to a neutralization tank where it is neutralised to pH 7.0 with soda ash. The mixture is passed through a filter press, the filtrate is decolorized by activated carbon and the clear filtrate concentrated in a triple effect evaporator. The liquid is again treated with carbon and concentrated in vacuum. The concentrated syrup (40 to 45°C) is quickly cooled and transferred to drums. The product contains 43% dextrose on a dry weight basis. The syrup can be used for various confectionery purposes, and after further purification is used for pharmaceutical purposes.

In the production of crystalline D-glucose monohydrate, the solution is evaporated under a vacuum to 70 to 88% solids, cooled to around 45°C and fed into 10,000 gallon crystallizers. The most common type of crystallizer is a horizontal cylindrical tank fitted with a cooling jacket and slowly rotating cooling coils. For seeding purposes, the product from a previous

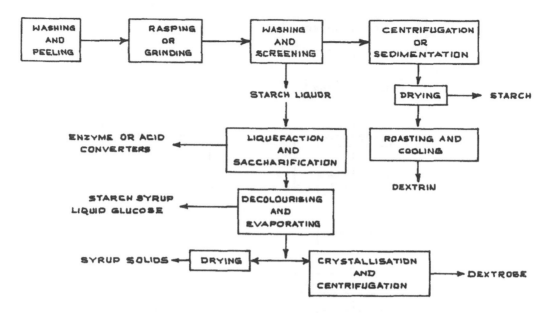

FIGURE 5.   Flow diagram for the production of starch and starch-derived products.

batch is left in the tank. After mixing the temperature is approximately 43°C. The mass is now slowly cooled according to present schedule to approximately 20 to 30°C over a period of 3 to 5 days, at the end of which time about 60% of the solids are crystallized as D-glucose monohydrate. The whole lot is fed into large centrifuge baskets lined with perforated screens. The product is collected and dried in large rotary dryers in a stream of warm air until it reaches a moisture content of 8%. The mother liquor is again decolored and crystallized to obtain a second lot of crystals. Recrystallized dextrose hydrate for use in intravenous injections, etc., is produced by redissolving the wet centrifuge cake followed by recrystallization. Anhydrous dextrose is obtained by crystallization of dextrose hydrate solution in vacuum, at 60-65°C.

In India, glucose syrup is used widely in the confectionery industry to provide sweetness and regulate crystallization. It is also used in combination with sucrose. It finds use in the pharmaceutical industry for use in intravenous feeding, for making tablets, and for energy foods. It is also used in foundries and as raw material for various microbial preparations.

IV. SAGO

Originally sago was derived from the palm, *Metroxylon* sp. found in Malaysia and Thailand. However, sago is now manufactured using tapioca starch. The initial steps are similar to starch production, i.e., peeling and washing, disintegration, and settling. The starch from the settling tanks is spread over cemented yards and partially dried in the sun. The partially dried material (40 to 45% moisture) is made into small granules by shaking in power-driven shakers or granulators which consist of wooden trays with cloth flooring. The oscillatory movement enables the starch granules to adhere together and form into globules. In the smaller units, the granulation is done manually by shaking small quantities (10 to 15 kg) of granules in cloth bags. The globules vary considerably in size and are sieved through standard sieves. According to IS-899-1956, 95% globules should pass through IS-sieve 170 but are retained on IS-sieve 85. The oversize granules are again powdered and granulated.

The next step is partial gelatinization. This is carried out by roasting the granules to gelatinize the surface. Wet globules are placed on shallow metal pans made of aluminium or iron which are smeared with small quantities of oil, and heated by fire. The granules are

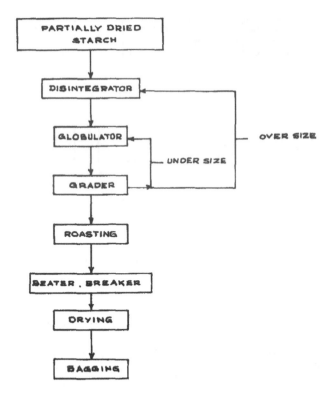

FIGURE 6.   *Flow diagram for the production of sago.*

stirred continuously throughout the operation for around 15 min. The granules are then dried in the sun or a hot-air oven (40 to 50°C). The dried mass is passed through a polisher to separate the large clumps into small granules. They are then graded according to size, color, and degree of roasting (Figure 6).

They are then packed in gunny bags (90 kg/bag). The yield of sago is around 25% of the weight of fresh tubers. Sago contains about 12% moisture, 0.2% protein, 0.2% fat, 87% carbohydrates, and has a calorific value of 351 calories per 100 g.

Sago is used mainly as infant and invalid food, and in preparation of puddings. Small-scale factories make pappad, vadam, etc. Over 1.5 metric tons of sago are produced yearly in India, mainly at Salem in Tamil Nadu.

V. CASSAVA MACARONI

Forms of macaroni can be classified into (1) extruded solid products, e.g., vermicelli, spaghetti, and noodles and (2) extruded hollow products, e.g, macaroni, elbows, tubes, rings, etc. The basic components are cassava flour, groundnut flour, and wheat semolina.

Cassava flour may be prepared from chips and should have a mesh size of 30 to 60, wheat semolina should be obtained from the best quality amber durum wheat, and should be free of flour and bran.

The basic procedure consists of making a dough which is cast through a die. Various shapes are obtained, and they are dried to less than 12% moisture content. Usually the plants contain 2 parallel sections for production of long goods and short goods.

The flours are received pneumatically in bins in the flour feed unit. The bins are provided with dosers, suitable filters, and automatic level indicators. The flours and semolina are fed into the mixer by dosers with preset value and mixed thoroughly. The flours are properly mixed with a little water to obtain a homogeneous dough. It is done under a vacuum. The dough is taken to a heavy speed duty press made of stainless steel with provided accessories. By compressing and forcing the dough through the dyes, the products are obtained. In case of long cuts, the products leaving the dyes are first hardened by a draught of warm air and then evenly spread on long sticks. The threads fold backwards at a preset time and are cut by a revolving blade. The unit is called a spreader. The long-cut macaroni spread on sticks are taken to a predryer, where the moisture level is brought down from 30 to 20%. The predried long-cut products are taken to a final dryer, where moisture content is reduced to 12.5%. The dried product is taken to a cutter unit, and cut into desired lengths. The stripped products are then recirculated to spreader. The product is taken to a hopper and packed.

In the case of short-cut macaroni, the flour mixing, dough formation, and processing are similar to long-cut macaroni. However, spreading operations are slightly different. Macaroni coming out of the extruders is very soft. It is hardened by subjecting it to a stream of warm air for a few minutes in a wire shaking screen (shaker). From the shaker, it is taken to a predryer where moisture content is reduced from 30 to 28%, by a preliminary warm air treatment. In the second stage, the moisture content is reduced to 18 to 20%, at 50-60°C for 45 to 50 min. The product is then taken to a final dryer, where the moisture content is reduced to 12 to 12.5%. It is then stabilized and packed.

A typical analysis of macaroni will be moisture — 10.8%, protein — 11.3%, carbohydrates — 76.0%, and fat — 1.8%.

VI. FRUCTOSE SYRUP

Fructose syrup has gained importance in view of the fluctuating prices of sugars and the harmful effects of synthetic sweetners. Fructose is 1.7 × sweeter than sucrose and 4 × sweeter than glucose. The conversion of glucose to fructose can be achieved by alkali or by an enzyme called glucoisomerase. Whereas an alkali gives only low levels of isomerization, isomerase can bring about a 40 to 45% glucose to fructose conversion. The conversion of starch is the same as for production of liquid glucose. The next step involved is the decolorization of the syrup which is achieved by either activated carbon treatment or passing through ion exchange columns. The decolorized sample is then subjected to isomerization with glucose isomerase in glass-lined tanks. The optimum temperature for the isomerization is 62°C and the reaction is carried out for 6 hr. The pH is maintained at 8.0. Stirring is carried out throughout the reaction. Samples are taken at regular intervals to check for the conversion by estimation of fructose content. When there is no more increase in the fructose content, the solution is withdrawn and again decolorized by either carbon or ion exchange resins. The solution may be concentrated in vacuum to desired levels of solids content. Pure fructose crystals can be obtained by separation on ion exchange columns, concentration, and seeding out with crystals of pure fructose (Figure 7).

VII. DEXTRIN

A. Manufacture Process

Radley[22] has correctly pointed out that manufacture of dextrin is a simple, well-known process, more of a craft than a science, and more experience at it gives the best results. Hence, the myth that dextrin production is highly complicated, is not true.

The steps involved in the dextrin production depend on the type of product desired. However generally they can be catalogued (1) predrying (2) acidification and (3) converting as described by Evans and Wurzburg.[23]

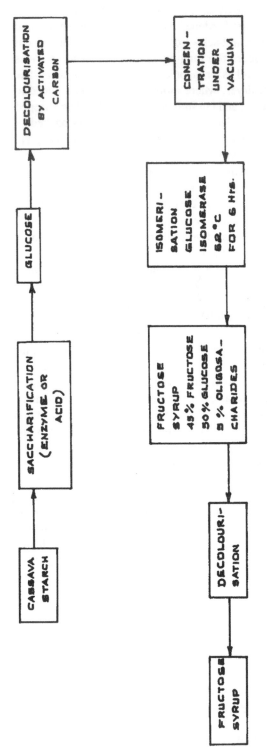

FIGURE 7.   Flow diagram for the production of high-fructose syrup.

1. Predrying

Moisture content of starting starch is to be brought to between 1 to 5% and hence starch may have to be subjected to drying. Moisture promotes hydrolytic reactions and is not desirable when products of high, cold-water solubility, and good solution stability are desired. Condensation reactions are usually negligible when moisture content is about 3%. The drying may be carried out in a dextrin convertor or a vacuum convertor, which can remove moisture much faster.

2. Acidification

Acidification is carried out by spraying a dilute solution of HCl into powdered starch containing less than 5% moisture. The starch may be tested for various factors like pH, acid factor, and ash content. In addition, the fiber content may also be determined for cassava starch. Depending on the properties, the starch or flour may be designated as "soft" or "hard". A soft flour starch is considered one that will convert to a given degree without much treatment. Cassava starch, which can dextrinize easily, would normally have a pH of 4.5 to 5.5, have a reasonably low acid factor, and an ash content below 0.1%.

The time of mixing, the homogeneity of the acidified mass, and the moisture content are crucial factors that decide the quality of dextrins. To ensure uniform distribution of the catalyst in the starch, a volatile acid is normally used. The acid (0.05 to 0.15%) is sprayed into starch which is kept agitated in a horizontal or vertical mixer. The acid usually used is hydrochloric acid, but nitric acid may also be employed. The amount of acid used may also depend on the source of the starch. For example, for potato dextrin, the acid used is 80 mℓ concentrated HCl in 300 mℓ water for 100 kg starch; for cassava, the acid to be used is 200 mℓ diluted to 4× its volume.[22] Various patents suggest use of gaseous hydrogen chloride. Oxidative catalysts such as chlorine give the products which are more stable to retrogradation. The reduced tendency to retrograde is brought about by formation of carboxyl groups by the oxidizing agent. The oxidation also helps improve clarity of the paste. Bulfer and Gapen[24] found that treatment with monochloroacetic acid followed by chlorine gas produces white dextrin of improved quality. Aluminium chloride has been found to reduce the conversion time. The catalyst is added to the starch slurry to an extent of 0.05 to 0.5% of the weight of starch at a pH 2.8 to 3.4. Calcium chloride may also be used.

Alkaline conditions may also be used for dextrin manufacture. Solution of sodium carbonate, ammonium hydroxide, and urea are blended with starch and the pH is maintained at the desired level. These reagents act as buffers, particularly in the production of British gums where acidity develops at high temperatures.

3. Conversion

The equipment in which this operation is carried out are large mixers mounted horizontally or vertically with capacities varying from 45 to 4500 kg of starch. Heating is carried out by steam, oil, or direct heating. The temperature increase and moisture content have to be strictly controlled in order to obtain proper conversion. If temperature increase is low in presence of high moisture content, a product with low viscosity and highly reduced sugar content is obtained. Often a stream of air is passed to remove the excess moisture. Care has to be taken to prevent possible charring of starch near the walls of the heating elements. Staerkle and Meier have suggested use of low pressure during dextrinization.[25]

When the starch is brought up to final temperatures of 95 to 120°C under specified acidic conditions, white dextrins are obtained and between 150 to 180°C, a series of low viscosity canary dextrins are produced. For the manufacture of British gum, the temperature used is 170 to 195°C and the time in the cooker is 7 hr or more.

The converted dextrin is transferred to another mixer for immediate cooling to avoid overconversion. The mixer is water jacketed to help cooling. After cooling, the dextrins may be neutralized using ammonia and packed.

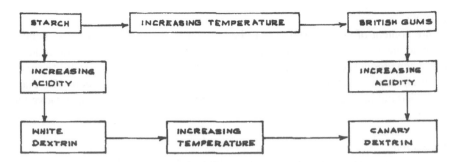

FIGURE 8.   Flow diagram for the production of dextrin.

B. Equipment Used in Dextrin Manufacture

A number of converters have been reported in the manufacture of dextrins. Usually bulk type converters with capacity of thousands of tonnes are used for the purpose. A horizontal or vertical gear driven ribbon agitator slowly stirs the acidified mass. Heating is done by using steam, hot oil, or gas combustion products and these are circulated through the side and bottom walls of the container. The main disadvantages of these type of converters are the lack of uniformity of temperature throughout the mass and the slow rate of heating. An improved system containing a vertical agitator was patented by Rowe and Hagen.[26] They achieve a more uniform transfer of heat by dividing the jacket into zones having both a steam inlet and outlet. Phillips[27] patented a continuous conversion process in which a thin layer of acidified starch is transported by a moving belt through heated ovens.

Staerkle and Meier[25] use a nonoxidizing environment during starch dextrinization, which helps to avoid discoloration. Reduced pressure, or atmosphere of nitrogen, carbon dioxide or sulfur dioxide, is used. The acid is applied to the starch in a vapor form after partial vacuum drying. Another advantage claimed for this process is that other chemicals may be added as desired and derivatization may be simultaneously done with dextrinization.

In another process developed by Ziegler et al.,[28] starch is passed by applying vibration at a frequency of 1250 per min and an amplitude 2 to 4 mm through a heated spiral tube conveyor. The dextrinized starch is discharged to a conveyor.

Conversion of starch in a fluidized bed has been done by Fredrickson.[29] Addition of 1% calcium phosphate is required to promote fluidization of starch. The advantages of this process are improved quality, product uniformity, and time saving. Other advantages are that the starch may be dried, acidified, dextrinized, and cooled all in one operation and that dextrin may be blended with other agents at the same time.

A plant for continuous manufacture of dextrin is in operation in Germany.[30] The starch is stored in silos from which it is transferred pneumatically to the acidifiers in which acid is sprayed through turbo nozzles. The acidified starch is mixed and pumped to a predrier, where the moisture content is reduced under vacuum to 2 to 3%. The predried starch is carried pneumatically to a kiln-type dextrinizer which is approximately 3 feet in diameter, 20 feet long, rotates at 3 to 12 rpm, and operates at a steam pressure of 8 to 10 atmospheres. The product after dextrinization is taken to a cooling tower and allowed to fall freely through cool air. The dextrin is mixed, sieved, and subjected to remoistening by passing through a tower where it comes into contact with a fog of moist air. It is then sieved and packed (Figure 8).

VIII. FERMENTED PRODUCTS FROM STARCH

A. Cassava Alcohol

The concept of production of chemical fuels like alcohol from plant biomass and its use

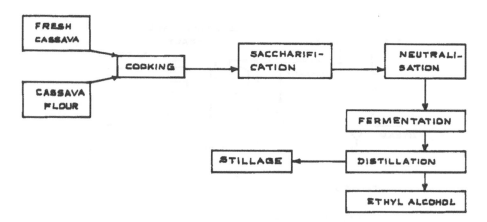

FIGURE 9. Flow diagram for the production of alcohol.

as energy source is an age old one. Photobiological energy conversion is a technique where carbon-based fuels can be obtained from plant sources, and in the frantic search for alternative renewable resources, certain sugar crops and starch crops like cassava could also be identified as suitable for this use. Although the income elasticity of cassava is considered to be low, and although in terms of ethanol production sugar crops like sugarcane do enjoy a better competitive position presently, in the future cassava can also become a strong contender for processing of ethanol. The ability of cassava to compete with sugar crops for ethanol production will largely depend on the total production cost, and this production cost cut considerably through increased use of cheaper raw materials. Studies conducted earlier showed that for ethanol production from sugarcane or cassava, approximately 35% of the final cost is accountable to production cost, and the remaining 65% is the cost of the raw materials alone. As cassava is a drought resistant hardy crop and has the ability to grow well even in poor and marginal lands that are generally considered unsuitable for other crops like sugarcane, it may in the long run enjoy a comparable position, if not better, as a raw material source for an alcohol fuel economy. Cassava with the possibility of its storage as chips, extends the scope of operating the distilleries for a much greater length of the year, versus the sugarcane-based distilleries which are mostly seasonal in operation. Both cassava and sugarcane can be combined advantageously in alcohol fuel and many sugar mills can be utilized for an integrated fuel economy.

1. Biotechnology of Cassava Alcohol Production

Production of ethanol from starch is not new in fermentation technology. Fresh cassava roots or cassava flour can be used for ethanol production. The first step is gelatinization and conversion of starch to simpler sugars by a process called saccharification which is accomplished with the help of saccharifying agents like mild acids, amylase enzymes, and substances containing amylase enzymes, e.g., malt. The hydrolysis process and the utilization of an efficient low-cost saccharifying agent are important factors in the production of fuel from starch (Figure 9).

The starch is cooked first to release the starch granules, which are bound to the ligno-cellulosic compounds of the roots. This will also facilitate the reaction between the saccharifying agents (acids, enzymes, etc.) and the lower substrate. Cassava starch, having a lower swelling and gelatinization temperature, can be easily saccharified to simple sugars with the help of acids or enzymes. The main advantage of cassava over any other energy crop is the presence of high fermentable sugars after saccharification. The use of a dilute acid solution helps to recover approximately 98.8% of the reducing sugars from the starch. The formation of a secondary reversion reaction which sets a limit to the yield of glucose

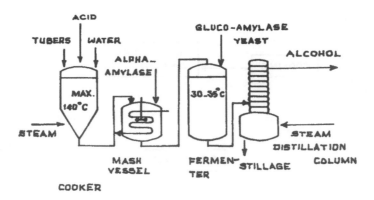

FIGURE 10.   Process diagram for alcohol production.

that may be obtained, the formation of high inorganic salts due to pH adjustment, and the corrosion of the machinery are some of the disadvantages of the use of acids in the saccharification process.

The enzyme amyloglucosidase (glucoamylase or glycoamylase) derived commercially from strains of the molds *Aspergillus* and *Rhizopous* can hydrolyze gelatinized starch completely to glucose. It acts by cleaving the α-1, 4-glycosidic linkage of the nonreducing terminal glucose unit. It can also hydrolyze the α-1, 6 linkage from the amylopectin. Malt contains the three most important enzymes for starch breakdown, namely, α-amylase, β-amylase and amyloglucosidase. The α-Amylase splits the α-1, 4 links randomly within the molecules, forming dextrins, which are small chains of glucose. This makes the gelatinized starch slurry more fluid and supplies more chain ends for the action of saccharifying enzymes. Since all the α-amylase cannot split the α-1, 6 links of starch and dextrins, all the branch points remain intact after α-amylase action. The amylase also breaks the α-1, 4 links of dextrin and starch, but only from the nonreducing ends of the molecules, resulting in maltose formation. Since neither of these enzymes attack the α-1, 6 linkages, their combined action converts only up to 85% of the starch into reducing sugars. Amyloglucosidase splits the α-1, 6 linkage of the maltose and short-chain carbohydrates, thus completing the hydrolysis of starch into fermentable sugars.[31] The continued action of the above enzymes produced from microorganisms can also split the cassava starch to glucose units.

Since amylase enzymes do not tolerate temperatures above 55°C, they must be added after gelatinization has taken place. The use of heat-stable bacterial amylase has received considerable attention at least for effective conversion of starch to sugars.

Large volumes of the saccharified starch are fed into fermentation vessels and inoculated with actively growing yeast, *Saccharomyces cerevisiae*. Many strains of yeasts, with varying capabilities and tolerances, have been reported. The yeast inoculation is usually 5 to 10% of the total volume and is grown aerobically in stages, from a laboratory pure culture. The optimum concentration of sugars for ethanol fermentation is 12 to 18%. The pH of the mash for fermentation is optimally 4 to 4.5 and the temperature of fermentation is 28 to 32°C. Alcohol is recovered from the fermented mash after 48 to 72 hr. At the end of fermentation, the yeast is separated from the mash by centrifugation or sedimentation and used for the next batch of fermentation. The resulting liquid is distilled for the recovery of alcohol (Figure 10).

2. Biochemistry of Alcohol Fermentation

Even in the absence of an external electron acceptor, many organisms can still oxidize some organic compounds with the release of energy by utilizing a process called anaerobic fermentation. The conversion of simple hexose sugars to ethyl alcohol by yeasts through a well-documented series of enzyme catalyzed reactions was elucidated by the classical studies of Mayerhof and Parnas in 1930.

The overall reactions involved in the conversion of glucose to ethanol by yeasts in the absence of oxygen can be represented as:

$$C_6 H_{12} O_6 \rightarrow 2 CH_3 CH_2 OH$$

Glucose   Ethanol

+

$2 CO_2$

Carbon dioxide

+

57 Kcal.

Energy

3. Byproducts from the Starch Alcohol Industry

Large quantities of organic wastes could be expected from cassava distilleries. These wastes could be effectively disposed of by converting them to poultry or cattle feed after enrichment and proper drying. Approximately 150 kg of waste material containing cellulose and sugar could be expected from 1 metric ton of cassava flour after fermentation.

4. Problems Associated with Fermentation

The conventional batch fermentation requires high cost manpower, because permanent supervision is required and the preparation procedure is elaborate and has to be repeated within small intervals of time. The shorter time of fermentation reduces the danger of contamination, but requires an uneconomic interruption of the process. The low production of ethanol is one of the serious disadvantages of batch fermentation. The yields of batch fermentation which is mostly performed in alcohol distilleries are about 6% by volume for a fermentation time of 36 hr. The productivity may be explained by the fact that the physiological properties of the microorganisms are not sufficiently considered in batch fermentation. The result is a low propagation rate of the cells. High substrate concentration at the beginning and the increasing product concentration at the end of the fermentation process cause inhibition of growth.

5. New Trends in Bioethanol Technology

The problems of substrate inhibition and product inhibition could be solved by adopting new trends in bioethanol technology.

a. Continuous Fermentation

This is characterized by continuous addition of nutritive media and by continuous removal of cells and fermenter solution. The initial growth of the cells is identical with that in batch fermentation. When a favorable exponential growth phase is obtained, the organisms may be kept permanently in this stationary phase by regulation of influx and efflux. The whole system is thus in a steady state. The yeasts are kept for a very long time in an exponential growth phase; a higher substrate transformation rate may be obtained in a very short time. As a result of this increased productivity, even smaller plants are economical. The danger of contamination and the possibility of mutation of the organisms are some of the problems encountered in continuous fermentation.

Table 4
PRODUCTIVITY OF ALCOHOL FROM CASSAVA, SUGARCANE, AND MOLASSES

Crop	Crop yield[a] (t/ha)	Fermentables (%)	Yield of alcohol (ℓ/t)	Alcohol productivity (ℓ/ha)
Cassava	16.9 (raw tubers)	30[b]	150	2490
Sugarcane	56	9.6	48	2688
Molasses	2.8[c]	50	225	686

[a]   8- to 10-month-old rainfed crop (Indian average).
[b]   From varieties containing 30% starch.
[d]   Byproduct of sugar industry.

b. Vacuum Fermentation

Vacuum fermentation is done in a simple stirred tank fermenter. In producing a vacuum (32 to 35 mm/kg) in the bioreactor, a mixture of ethanol and water is distilled directly at 30°C from the fermenter. The yeast cells are not hampered by this procedure; in contrast, the product inhibition occurring at ethanol concentrations of maximum 6% in the medium is overcome. The technical application of vacuum fermentation, however, has become questionable because of the high energy requirements of the cooking and pumping system.

c. Immobilized Yeast Cells

The most recent advance in alcohol production technology is the application of immobilized cells in continuous column reactors. The approach combines several advantages: (1) high cell concentration can be obtained; (2) the reaction rate is accelerated; (3) operation can be performed at a higher dilution rate without a wash out; (4) the end product inhibition is eliminated, since alcohol is removed continuously; (5) anaerobic condition can be achieved, since cells are entrapped into gel matrix; and (6) costly equipment designs such as fermenters, agitators will be eliminated.

d. Genetic Improvement of Yeast Strains

It is also necessary to improve the yeast strains genetically for adaptation at higher substrate and end-product concentrations. Improvement of yeast strains for higher temperature tolerance is another area where immediate attention of microbiologists is required.

6. Energy Gain or Loss

In the agricultural phase, sugarcane uses the highest amount of energy; on the other hand, cassava utilizes twice as much labor. The industrial phase is always more energy intensive, consuming 60 to 75% of total energy. When transferred to a common scale, the energy requirement for sugarcane and cassava is 4,138 and 2,575 Mcal/ha/year, respectively.[32] It is true that the processing of cassava requires steam for liquefaction and hydrolysis. It is estimated that approximately 100 kg of steam is required for producing 100 ℓ of alcohol from cassava and a net energy ratio of 4.7 has been reported.[32]

It can be seen that cassava has a higher percentage of fermentables than sugarcane and the yield of alcohol from it per tonne of raw material is higher (Table 4). This alone will make the price of power alcohol from cassava more attractive than that from sugarcane. Rapid multiplication of the high yielding and high starch-containing varieties can lead to a higher alcohol yield per hectare of cassava, further bringing down the cost. It should be remembered that the price of alcohol from molasses has become attractive only because of the statutory control imposed on the price of molasses.

It is considered that alcohol fuel can be the most immediate substitute for the expensive petroleum-based fuels. It is also well-known that in some aspects ethyl alcohol is a better fuel than gasoline and has got certain favorable qualities as an internal combustion engine fuel. In Brazil, ethyl alcohol was considered to be the most promising nonfossil and renewable fuel and already a good amount of basic information has been generated on its practical utility as an engine fuel. Ethyl alcohol is mixable with gasoline and blends up to 20% alcohol with petrol are possible, without any basic modification of the gasoline engine. Cassava starch alcohol can thus act as a good liquid energy as a gasoline replacement, although it need not necessarily be a cheap proposition. In the overall agro-energy production system, cassava can play a very vital role to produce liquid energy.

Studies conducted in various countries revealed the benefits of agricultural energy production system using cassava as the raw material. Emphasis has been put on the great potentiality of small scale cassava alcohol units for the less developed tropical countries, and certain designs of small equipment suitable for village-based cassava alcohol industry have already been suggested. Based on the reports from Brazil, every million liters of cassava ethanol resulted in 96 to 280 additional jobs and required 320 to 660 ha of land. So, in many of the developing tropical countries where sufficient unused land is available for the production of crops like cassava, such energy farming may help in generating sufficient employment opportunities and boost the rural economy.

B. Lactic Acid Fermentation

Several carbohydrates such as cassava starch, potato starch, whey, and molasses could be used for the production of lactic acid. The starch has to be saccharified into sugars before fermentation. The techniques of saccharification of starch to sugars have been discussed in detail previously. The organisms *Lactobacillus plantarum*, *L. leichmania*, *L. mesenteroides*, and *L. delbruickii* are the different types of bacteria capable of fermenting sugars for lactic acid production.

The sugar solution thus obtained is suitably diluted to the desired sugar concentration in the fermenters and cooled to ambient temperatures using plate heat exchanger. After the addition of nutrients, the pH is adjusted to 6.5 to 7.0 and the lactobacillus is inoculated. Fermentation is carried out at room temperature and during the process, calcium hydroxide is added intermittently to neutralize the acid and also calcium lactate is formed. Upon completion of fermentation (2 days), the material in the fermentation unit is boiled to coagulate the protein which is then filtered. The filtrate containing the calcium lactate is then concentrated by removal of water under vacuum, followed by additional treatment for purification of the compound.

C. Acetone-Butenol Fermentation

The acetone butenol fermentation is an anaerobic fermentation brought about by *Clostridium acetobutylicum* and closely related species or variants. The important products produced by these organisms are butyl alcohol, acetone, and ethyl alcohol.

The primary carbohydrate raw materials for the butenol-acetone fermentation processes are corn, cornblack strap molasses, and high test molasses. A number of other raw materials including cassava starch have been tried and successfully used for the acetone butenol fermentation. Gelatinized starch with the addition of nitrogenous and phosphatic nutrients can serve as an excellent medium for the acetone-butenol fermentation.

Ammonia may be supplied in the form of ammonium sulfate or as aqueous ammonium hydroxide. If ammonium sulfate is used a buffer such as calcium carbonate may be employed to neutralize the sulfuric acid remaining, as the ammonium ion is metabolized. Generally phosphate is applied as superphosphate.

The temperature should be maintained at 35° to 37°C and after a fermentation period of

16 hr with *Clostridium acetobutylicum*, significant increase in the ratio of butenol to other solvents resulted. Temperature control generally involves cooling since heat is evolved during fermentation. Outside water curtains, i.e., water sprays on the exterior of the fermenters have been commonly used for cooling. When 10 to 12 t of cassava was charged into each fermenter holding 150 ℓ of mash containing optimum sugar concentration of 5%, it yielded 30 to 34% of solvents. Vegetable oil cakes such as soybean cake, cottonseed cake, rapeseed cake, and peanut cake as well as rice bran have been successfully used as nutrient materials in acetone-butenol fermentation.

D. Itaconic Acid Fermentation

The possibilities of the production of itaconic acid, an unsaturated dicarboxylic acid, from cassava starch was successfully demonstrated by the scientists of Central Tuber Crops Research Institute.[34] Three-days-old spore suspensions of *Aspergillus itaconicus* (NRRL 1960) when inoculated with a cassava starch medium containing 4% cassava starch, 0.25% NaCl. The pH of the medium has to be adjusted to 4.5. Two mℓ of concentrated rice bran liquor made by boiling 500 g rice bran soaked in 100 mℓ of water for 30 min was added to the medium.

The spores of *Aspergillus itaconicus* harvested from the growth medium after 3 days of incubation at room temperature were used for inoculating the production medium. The concentration of the inoculum was in the ratio of 1:10. After 15 days of incubation at room temperature, the fermented liquor was concentrated under a vacuum to one tenth of its original volume. The itaconic acid crystalized from the hot concentrated liquor and was filtered and dried. The percentage of itaconic acid recovered in cassava starch was 34.5 vs. 37.1 in sucrose medium. Cassava starch could thus be utilized as a substitute for sugar in the production of itaconic acid as a carbon source.

IX. SILK PRODUCTION FROM CASSAVA

Silk has been the undisputed queen of textiles for centuries, though it comprises only a mere 0.2% of the total production of textile fibers of the world. The silk fibers are one of natures gifts to mankind, and is produced in nature by certain caterpillars which belong to two families: (1) Bombycidae (the commercial silkworm) and (2) Saturnidae (wild silkworm) of the order lepidoptera (butterflies and moths).

The most common source of silk, the *Bombyx mori* is raised domestically, but only where there are mulberry leaves to satisfy its finicky appetite. More robust than their domesticated cousins, wild silkworms produce a tougher, rougher silk not as easily bleached and dyed as the mulberry silk. The Eri silkworm, *Philosamia ricini*, raised on the castor plant in India, produces silk that is extremely durable but that cannot be easily reeled off the cocoon and must be spun like cotton or wool. Utilization of cassava leaves for the production of Eri silk is a practice in some tracts of Northeastern India. Though the silk produced by rearing the Eri silkworm is not of high quality, in areas where cassava is grown, this could be attempted without much additional effort (Figure 11). Cassava being an annual crop, a period of at least 11 months lapses before the returns on investment accrue to the farmer.

The Eri silkworm has a life cycle of 45 days. An adult female silkworm lays about 300 to 500 eggs in clusters on "Kharikas" (sticks). A vertical position is preferred for oviposition. Oviposition is confined to nighttime and continues for 2 to 3 days.

After hatching, the larvae could be fed on cassava leaves for about 3 days. Then they stop feeding and settle for first moult (casting off the skin). Moulting time is about 24 hours. During the feeding time fresh leaves have to be given 3 to 4 times a day, depending on the appetite of the worm.

FIGURE 11. An Indian woman wearing an Eri silk sari.

After first molt, the larvae start feeding again for 2 to $2^{1}/_{2}$ days and prepare for the second molt of 1 day duration. Third instar larvae feed for about 3 days and then undergo the 3rd molt. The duration of the fourth larvae step is about 3 days with a molting time of 24 hr. The fifth instar larvac feed for 4 to 5 days, then pupate.

Fully matured larvae which will be on the move in search of a place to pupate are transferred to a basket containing dry leaves or straw. They spin their cocoons in the night between the dry leaves provided. Spinning will be finished in 3 to 4 days and on the 6th day the cocoons are collected.

Moths of Eri silkworms are allowed to emerge from the cocoons, which are then cleaned, and the pupae shaken out along with the last skin of the worm. Eri cocoons are elongated, open at one end, and do not form a continuous filament. Therefore, these cocoons are spun and not used for reeling. Cocoons may be dull white or brick red in color. They are washed, boiled for 2 to 3 hr in alkaline water and left wrapped in green leaves for 3 to 4 days. After this they are washed in pure water, opened out under the water in the form of discs, and stored until required for spinning. Eri silk is usually handspun.

The Eri silkworms are voracious leaf-eating caterpillars and usually feed on caster leaves but also do well on cassava leaves (Figure 12). The caterpillars are reared indoors on trays supplied with fresh leaves and 5 to 6 generations can be raised in a year. Initially they are fed 3 times a day and 5 times a day during the last stage. Leaves are frequently changed and fresh leaves are given. Defoliation of cassava leaves for rearing Eri silkworms has to be done with the utmost care as the increased rate of defoliation may effect the tuber yield. Leaf supply has to be maintained during the day and at night. A quantity of 15 kg fresh leaves spread over a period of 20 days is required to produce 1 kg of cocoons.

FIGURE 12.   Eri silkworms feeding on cassava leaves.

The nitrogen balance of Eri silkworms reared on cassava leaves of the variety ME 116 was determined and it was found that for the formation of silk and other biochemical changes there was a reduction of 9677.6 g nitrogen per hectare from the final larval stage. From the pupal to moth stage and from moth to egg laying stage the nitrogen reduction was 5760.7 and 4479.1 g/ha respectively.[35] Dried cassava leaves stored for 9 months at room temperature were found to give similar general performance of the silkworms as that of fresh cassava leaves.[36] The problems associated with leaf defoliation and reduction in yield could be solved to a great extent by rearing the worms on powdered, dried leaves collected from the farms.

Eri culture can serve as an alternate source of employment for farm families. For a large class of farm families, growing cassava with a holding size of around 0.2 ha is inadequate either to generate sufficient income for the family or to fully engage the labor of an average family, and thus Eri culture could be recommended. Male and female labor in these families are usually unemployed for want of suitable opportunities and also due to the sociological backgrounds inhibiting them to work as agricultural labor. An integrated approach for cassava production and Eri culture, if properly planned and executed, can generate additional employment and income for the farm families.

REFERENCES

1. **Jones, S. F.,** *The world market for starch and starch products with particular reference to cassava (tapioca) starch,* G173, Tropical Development and Research Institute, London, 1983, 98.
2. **Anon,** *Cassava: export potential and market requirements,* International Trade Centre UNCTAD/GATT, 1977, 65.
3. **Radley, J. A.,** *Starch Production Technology,* Applied Science Publishers, London, 1976, 587.
4. **Grace, M. R.,** *Cassava Processing,* Plant Production and Protection series No. 3, FAO, Rome, 1977, 155.
5. **Edwards, D.,** *The Industrial Manufacture of Cassava Products: An Economic Study,* G88, Tropical Products Institute, London, 1974, 43.
6. **Caransa, A.,** Simplified economical cassava starch process, *Starke,* 32, 48, 1980.
7. **Buckle, T. S., Zapata, M., L.E., Cardenas, O. S., and Cabra, E.,** Small-scale production of sweet and sour starch in Colombia, in *Cassava Harvesting and Processing:* Proc. of a workshop held at CIAT, Cali, Colombia, Webber, E. J., Cock, J. H., and Chouinard, A., Eds., IDRC, Ottawa, 1978, 26.
8. **Green, D. W. and Maloney, I. O.,** *Perry's Chemical Engineers' Handbook,* McGraw-Hill, New York, 1984, Chaps. 8 and 19.
9. **Birse, D. G. and Cecil, J. E.,** *Starch Extraction: a Checklist of Commercially Available Machinery,* G142, Tropical Products Institute, London, 1980, 16.
10. **McCann, D. J. and Saddler, H. D. W.,** A cassava based agro-industrial complex. Part 1: Financial considerations, *Proc. Chem. Eng.,* 28, 230, 1975.
11. **Heltne, L. P.,** Improved process and apparatus for the treatment of dried cassava roots to obtain starch, British Patent, 468, 926, 1937, 4.
12. **Koch, L. A.,** Een globale bepalling van het zetmeelgehalte in geschilde cassave wortels voor proefdoeleinden. B. Uitkomsten van een vergelijkende proef met 15 cassava varietein, (A. a global method for determining the starch content of peeled cassava roots for experimental purposes. B. Results of a comparative trial with 15 cassava varieties), Dutch, cited in *Abstracts on Cassava,* CIAT, Cali, 5, 56, 1979.
13. **Wholey, D. W. and Booth, R. H.,** A comparison of simple methods for estimating starch content of cassava roots, *J. Sci. Food Agric.,* 30, 158, 1979.
14. **Anon.,** CIAT Annual Report 1975, Centero Internacional de Agricultura Tropical, Cali, 1975, B-41.
15. **Krochmal, A. and Kilbride, B.,** An inexpensive laboratory method for cassava starch extraction. *J. Agric.* (University of Puerto Rico), 50, 252, 1966.
16. **Ramanujam, T.,** Extractable starch in six released varieties of cassava and its relationship with dry matter, in *Proc. Natl. Symp. Production and Utilization of Tropical Tuber Crops,* Abstracts, Indian Society for Root Crops, Trivandrum, 1985, 19.
17. **Thanh, N. C. and Wee, J. S.,** Treatment of tapioca starch waste waters by Torula yeast, *Can. Inst. Food. Sci. Technol.,* J., 8, 202, 1975.
18. **Prescod, M. B. and Thanh, N. C.,** Treatment alternatives for waste waters from the tapioca starch industry, *Prog. Wat. Tech.,* 9, 563, 1977.
19. **Balagopalan, C., Maini, S. B., and Hrishi, N.,** Microbiological treatment of starch factory effluents and the production of single cell protein, *J. Root Crops,* 3, 4, 7, 1977.
20. **Manilal, V. B., Balagopalan, C., and Narayanan, C.,** Physico-chemical and microbiological characteristics of cassava starch factory effluents, *J. Root Crops,* 8, 27, 1983.
21. **Kunhi, A. M., Ghilolyal, N. P., Losane, B. K., Ahmed, S. Y., and Natarajan, C. P.,** Studies on production of alcohol from saccharified waste residue from cassava starch processing industries, *Starke,* 33, 275, 1981.
22. **Radley, J. A.,** Dextrin and British gums, in *Starch and Its Derivatives,* Vol. 2, John Wiley & Sons, New York, 1954, 108.
23. **Evans, R. B. and Werzburg, O. B.,** Production and use of starch dextrins, in *Starch — Chemistry and Technology,* Vol. 2, Whitler, R. L. and Paschall, E. F., Eds., Academic Press, New York, 1967, Chap. 11.
24. **Bulfer, A. J. and Gapen, C. C.,** Monochloroacetic acid and chlorine in dextrin manufacture, U.S. Patent 2,287,599, 1942.
25. **Staerkle, M. A. and Meier, E.,** Dextrinisation under low pressure and in presence of Cases, U.S. Patent 2,698,818, 1944.
26. **Rowe, W. J. and Hagen, C.,** Improved conventor for dextrins, U.S. Patent 2,332, 345, 1943.
27. **Phillips, N. C.,** Starch conversion products, U.S. Patent 1,994,570, 1933.
28. **Ziegler, C., Kohler, R., and Ruggeberg, H.,** Conversion of high molecular weight carbohydrate, U.S. Patent 2,818,357, 1957.
29. **Fredrickson, R. E. C.,** Dextrinisation of starch, U.S. Patent 2,845,368, 1958.
30. **Goerner, A.,** A new continuous dextrinisation process, *Starke,* 12, 365, 1960.

31. **De Menezes, T. J. B.,** Saccharification of cassava for ethyl alcohol production, *Process Biochem.*, 13, 24, 1978.
32. **Sambeays, M.,** Alcohol from cassava in Brazil. Pests threaten yields, *World Crops*, 31, 181, 1979.
33. **McCann, D. J. and Prince, R. G. H.,** Agro-industrial systems for ethanol production, in *Proc. Conf. Alcohol Fuels*, Sydney, 1978.
34. **Potty, V. P., Maini, S. B., Moorthy, S. N., and Balagopalan, C.,** Production of itaconic acid from Cassava starch, in *Proc. Seminar on Post Harvest Technology of Cassava*, Trivandrum, India, 1980, 31.
35. **Kallemurrahman, M. and Raj, S. P.,** Nitrogen balance of Eri silkworm (*Philosamia cynthia ricini* boisduval) reared on tapioca (*Manihot utilissima* pohl) leaves, *Proc Ind. Nat. Sci. Acad.*, 48, 55, 1982.
36. **Liu, Z. C., Wang, K. Y., Wu, C. T., Sun, Y. Y., Wang, C. Y., Chen, T. C., Tseung, C. C., Lian, Y. F., and Tam, L. S.,** Cria del hospedante artificial de la avispa tricogramatida (gusano de seda) con dietas artificiales, *Acta Entomol. Sinica*, 26, 165, 1983.

Chapter 11

ANALYTICAL METHODS FOR CASSAVA

The chemical constituents of cassava crops often decide the food and industrial value of the produce. For instance, the higher the starch content of cassava tubers, the higher price they demand from starch extraction units. Similarly, the lower hydrocyanic acid (HCN) containing varieties are preferred for cooking purpose. In this chapter, an attempt is made to compile various analytical techniques for the estimation of various biochemical constituents in cassava.

I. ESTIMATION OF STARCH IN TUBERS[1]

Five grams of fresh material are weighed in a 100 mℓ conical flask containing 25 mℓ of 80% ethanol. The flask is left overnight and filtered using Whatman No. 1 filter paper. The residue is washed with distilled water twice and transferred proportionally into a 100 mℓ conical flask, 20 mℓ 2 N HCl added and hydrolyzed by heating for 20 min in a water bath. Care is taken to avoid excessive foaming, and the volume is kept constant by the addition of distilled water, as necessary. The completion of hydrolysis is measured by the absence of blue color with N/10 iodine solution. The hydrolyzed solution is increased to 100 mℓ and after further dilution to obtain approximately 100 ppm, the glucose is estimated by any of the standard methods for sugar estimation.

II. FRACTIONATION OF AMYLOSE AND AMYLOPECTIN[2]

A. Extraction of Starch from Tubers

Fresh material is always used for starch extraction, especially for cassava, which should be processed preferably on the day of harvest. The skin and rind are removed, the tubers are washed free of dust and external particles and sliced into small pieces. The tissue is homogenized in a mortar or blender with the addition of a small quantity of an amylase inhibitor (usually 0.01 M mercuric chloride). The suspension is filtered through muslin or cheese cloth, and the homogenate (filtrate) is centrifuged at 200 to 300 × g. Repeated addition of water and centrifugation removes crude impurities. Alternately, the starch granules may be allowed to settle by keeping them undisturbed in the cold, decanting the supernatant, and resuspending the caked starch in a fresh quantity of water which is then allowed to settle again. Finally, the cake is filtered under suction, washed with alcohol or acetone, and oven-dried at a low temperature of 40 to 50°C.

B. Fractionation of Amylose and Amylopectin

Various procedures are available for separation of amylose and amylopectin from starch and the most common methods are given below.

1. Reagents

1. 0.157 N sodium hydroxide solution
2. 5% NaCl solution
3. 1 N HCl

2. Procedure

Starch (100 g) suspended in 1 ℓ of water is added, while gently stirring, to 20 ℓ of sodium hydroxide solution. The mixture is stirred until the solution becomes clear. This stirring should not be too vigorous, as it can damage the gel structure of the amylopectin. After the solution has become clear, it is allowed to stand for 5 min. Then 4 ℓ of NaCl solution is added and the solution is neutralized to a pH of 7.0 using HCl. The solution is allowed to stand for a 12 to 16 hr period. When the gel settles and a clear separating division between the gel and supernatant solution becomes visible, the supernatant is removed by suction or siphoning.

The amylose from the supernatant is separated from the solution by complexing with 1-butanol. The filtrate is saturated with distilled 1-butanol and stirred gently for about 1 hr at room temperature. The precipitated amylosebutanol complex is allowed to settle for 2 to 3 hrs, and the supernatant is siphoned off. The solid is then centrifuged at 3000 rpm for 15 min and the process repeated to obtain a purer sample. The butanol complex can be stored for a few days and the butanol can then be removed by bubbling in oxygen free nitrogen in a boiling water bath. The yield of amylose is approximately 10% of the total starch.

The amylopectin gel which had settled is centrifuged at 8000 rpm for 20 min at 20°C, the supernatant is discarded, and 1% NaCl solution is added (2 ℓ) and stirred vigorously. The gel is recentrifuged at 3000 rpm. After further washing with a brine solution and recentrifugation, the gel can be freeze dried or precipitated in excess alcohol. Approximately 40 g amylopectin may be obtained from 100 g of dry starch.

III. DETERMINATION OF AMYLOSE IN STARCH[3]

The method is based on the absorption of iodine by amylose to produce a blue color.

A. Reagents

1.   1 N Sodium hydroxide
2.   1 N and 0.1 N Hydrochloric acid
3.   Iodine solution, 0.2% in 2% KI

Dissolve exactly 100 mg of amylose in 10 mℓ sodium hydroxide by shaking, dilute to approximately 50 mℓ with water, add approximately three fourths the calculated amount of 1 N HCl required to neutralize the solution, and make up to 100 mℓ and store in refrigerator.

B. Procedure

Weigh exactly 100 mg of the starch into a 100 mℓ volumetric flask. Add 1 mℓ distilled alcohol and mix thoroughly. Introduce 10 mℓ sodium hydroxide solution and leave overnight. There should be no lumps or clots after this period. It is then diluted to 100 mℓ. It can be left overnight to dissolve completely. Pipette out 5 mℓ into a 100 mℓ volumetric flask. Add 3 drops of phenolphthalin indicator solution and 50 mℓ water. Add the diluted acid by drops and shake until the pink color disappears. Add 2 mℓ iodine solution, shake, and increase to 100 mℓ. Read at 630 nm using a standard amylose solution and iodine blank. For standard amylose, 1 mℓ of the stock solution may be treated as above.

$$\text{Amylose content} \ - \ \frac{\text{Unknown reading}}{\text{Standard reading}} \times \frac{\text{mg dry amylose (m}\ell \times 100)}{5 \times \text{dry solid in sample (g/g)}}$$

IV. ESTIMATION OF TOTAL SOLUBLE SUGARS[4]

A. Anthrone Method

The anthrone reaction is the basis of a rapid and convenient method for the determination of hexoses, aldopentoses, and hexuronic acid, either free or present in polysaccharides. The blue-green solution shows an absorption maximum at 620 nm.

1. Reagents

Dissolve 2 g of anthrone in 1 ℓ of concentrated H_2SO_4. Prepare fresh.

2. Procedure

A 0.5 g fresh sample or 100 mg of the dried sample is extracted with hot 80% ethanol for 30 min, centrifuged, and the supernatant is collected. The extraction is repeated, and the supernatant is pooled, and increased to 100 mℓ.

A suitable aliquot is taken and the alcohol is evaporated by keeping it in a water bath. Then 4 mℓ of anthrone reagent is added and allowed to overflow down the side of the test tube, then the tubes are placed in a boiling water bath for 10 min. Glass marbles are placed on top of the tube to prevent evaporation, the solution removed from the water bath, and cooled to room temperature. The green color developed is measured at 625 nm against the blank. A standard curve of glucose is prepared (20 to 100 μg) and the sugar content calculated.

V. ESTIMATION OF REDUCING SUGARS[5]

A. Nelson's Method

The sugar is heated with an alkaline solution of copper tartrate, and cuprous oxide is produced which reacts with arsenomolybdate to give molybdenum blue, and the color is then measured in the colorimeter. Sodium sulfate is included in the reaction mixture to minimize the entry of atmospheric oxygen into the solution which would cause reoxidation of the cuprous oxide.

1. Reagents
a. Copper Reagent A

Dissolve 25 g of anhydrous sodium carbonate, 25 g sodium potassium tartrate, 20 g sodium bicarbonate, and 200 g anhydrous sodium sulfate in 1 ℓ distilled water. Filter if necessary.

b. Copper Reagent B

Use 15% $CuSo_4 \cdot 7 H_2O$ containing 1 or 2 drops concentrated H_2SO_4/100 mℓ.

c. Arsenomolybdate Reagent

Dissolve 25 g ammonium molybdate in 450 mℓ of distilled water, add 21 mℓ of concentrated H_2SO_4. Add 3 g of $Na_2HASO_4 \cdot 7 H_2O$ dissolved in 25 mℓ of water, mix, and place in an incubator at 37°C for 24 to 48 hr. Store in a dark bottle to protect from light.

2. Procedure

Pipette out an aliquot from the sugar extract (containing 10 to 15 μg of reducing sugar) and add 1 mℓ of a mixture (prepared on the day of use) of 25 parts of reagent A to 1 part of reagent B, and mix the solutions. Heat the tubes in a boiling water bath for 20 min. Cool the tubes under a running tap and add 1 mℓ of arsenomolybdate reagent to each tube. The color develops very rapidly. Dilute to 25 mℓ and read at 520 nm. Plot a standard curve with the known concentration of glucose.

VI. CYANIDE ESTIMATION

Cassava contains two cyanoglucosides, linamarin and lotaustralin, which on hydrolysis produce hydrocyanic acid (HCN). Methods for determination of cyanide in cassava involve hydrolysis of the cyanoglucosides with endogenous or exogenous linamarase and estimation of the cyanide formed.

A. Alkaline Picrate Paper Method[6]

1. Reagents

For alkaline sodium picrate solution — Take 25 g Na_2CO_3 and 5 g picric acid and dissolve in 1 ℓ of distilled water.

2. Procedure

Take 1 g cassava tuber and homogenize in 25 mℓ water. Place in a 500 mℓ conical flask. Fix the cork with a Whatman No. 1 strip (20 cm × 2 cm) soaked in an alkaline sodium picrate solution. Keep the flask for 18 hr at room temperature. Remove strip and elute it in 60 mℓ water. Read the absorbance at 540 nm or with a green filter.

3. Preparation of Standard Graph

Use different concentrations of KCN solution containing 5 to 50 μg cyanide in a 500 mℓ conical flask. Add 25 mℓ of 1 N HCl. Proceed as given for the estimation of HCN and prepare a standard graph.

B. Enzymatic Assays for Estimation of Cyanide in Cassava

1. Cooke's Method[7]

Reagents:

1. 0.1 M orthophosphoric acid (5.6 mℓ O. phosphoric acid per liter of distilled water)
2. 0.1 M phosphate buffer, pH 6.0
3. 0.2 M phosphate buffer, pH 7.0
4. 0.1 M acetate buffer, pH 5.5
5. 0.2 M NaOH (8 g NaOH per liter distilled water)
6. 5 mM linamarin (36 mg linamarin is dissolved in 30 mℓ 0.1 M phosphate buffer, pH 6.0)
7. Chloramine T (0.5 g dissolved in 100 mℓ distilled water)
8. Pyridine/pyrazolone reagent: 0.2 g bispyrazolone and 1.0 g of 3 methyl-1-phenyl-5-pyrazolone are dissolved in 200 mℓ pyridine (prepare fresh)
9. Potassium cyanide for calibrating absorbance values: the KCN is dried for 12 hr over concentrated sulfuric acid in a desiccator. Just prior to use, 125 mg is dissolved in 500 mℓ of 0.2 M NaOH in a volumetric flask and diluted 100 times in 0.1 M phosphate buffer, pH 6 to give a solution of 2.5 μg/mℓ
10. Ammonium sulfate: Extracts of crude enzyme are raised to 60% concentration of $(NH_4)_2SO_4$ per liter of extract at room temperature.

a. Preparation of Linamarase from Cassava Rind

Diced cassava peel (25 g) is homogenized in a blender for 3 min in 200 mℓ 0.1 M acetate buffer, pH 5.5. The homogenate is centrifuged at 10,000 g for 30 min and the supernatant is brought to 60% saturation of ammonium sulfate and held at 4°C for 16 hrs. The precipitate obtained by centrifugation at 10,000 g for 1 hr is dissolved in 25 mℓ 0.1 M phosphate buffer, pH 6.0, and dialyzed against this buffer. The enzyme solution should be stored at 10°C and maintains its activity for 2 to 3 months.

b. Determination of Linamarase Activity

The activity of the linamarase solution must be determined prior to its use in cyanide determinations. Enzyme aliquots (0.1 mℓ) are added to tubes containing 5 mM linamarin (0.5 mℓ). The tubes are incubated at 30°C for 30 min and the reaction is stopped by adding 0.6 mℓ of 0.2 N NaOH.

The cyanide present in each tube is determined spectrophotometrically as follows. A 2.8

mℓ of 0.1 M phosphate buffer, pH 6.0, is added to each tube, followed by 0.2 mℓ of chloramine T. The solutions are mixed and the tubes placed in the water for about 5 min. 0.8 mℓ of the pyridine/pyrazolone reagent is then added to each tube followed by a thorough mixing. The blue color that develops after 90 min at room temperature is measured at 620 nm.

KCN dried over concentrated H_2SO_4 is used to calibrate the absorbance values. The absorbance measured in the enzyme assay is used to calculate the activity in units per mℓ of enzyme solution, where one unit is defined as that which produces 1 μM of HCN per min at 30°C. Usually the enzyme has an activity of about 10 units per mℓ. The enzyme activity required to totally hydrolyze the cyanogenic glucosides present in the extracts is 0.3 units.

c. Preparation of Cassava Extracts for Cyanide Analysis

One cm cubes of parenchymal tissue (30 g) are sliced and homogenized at room temperature in a blender for 2 min in 160 mℓ of 0.1 M phosphoric acid. The resulting homogenate is filtered or centrifuged and increased to a known volume.

d. Determination of Cyanide in Extract

Aliquots (0.1 mℓ) are pipetted into stoppered test tubes containing 0.4 mℓ of 0.2 M phosphate buffer pH 7.0. 0.1 mℓ enzyme is added and the tubes are incubated at 30°C for 15 min and the reaction is stopped by adding 0.6 mℓ of 0.2 M NaOH. The cyanide present in the tubes is determined by the chloramine T pyridine/pyrazolone method described earlier.

Open tubes can be utilized instead of closed tubes, since the loss of cyanide is insignificant.

2. Spectrophotometric Method[8]
Reagents:

1. 80% ethanol — dilute 80 mℓ of 98% ethanol to 100 mℓ with distilled water
2. 0.1 M phosphate buffer pH 6.0
3. 0.2 M NaOH (8 g NaOH dissolved in 1 ℓ distilled water)
4. 0.2 N HCl (1 mℓ concentrated HCl is diluted to 60 mℓ with distilled water)
5. 1% Chloramine T (1 g chloramine T is dissolved in 100 mℓ water).
6. Barbituric acid/pyridine reagent 1.5 g barbituric acid is dissolved in a little water and 15 mℓ pyridine. Then 3 mℓ concentrated HCl is added and the solution is increased to 50 mℓ with distilled water.

a. Procedure
i. The Preparation of Linamarase from Cassava Rind is Similar to Cooke's Method
ii. Determination of Linamarase Activity

The reaction system is similar to that of method 1. Enzyme aliquots (0.1 mℓ) are added to tubes containing 5 mM linamarin (0.5 mℓ). The tubes are incubated at 30°C for 30 min and the reaction is stopped by adding 1 mℓ of 0.2 N NaOH, and 1 mℓ of 0.2 N HCl is then added to neutralize H, 1 mℓ of 1% chloramine T is added and after 1 min, 3 mℓ barbituric acid/pyridine reagent is added. The volume is increased to 25 mℓ and the absorbance of the pink color is measured at 570 nm after 10 min. Potassium cyanide (KCN) dried over concentrated H_2SO_4 is used to calibrate the absorbance values.

iii. Preparation of Cassava Extracts for Cyanide Analysis

Diced 1 cm cubes of cassava tuber (10 g) are extracted twice with 25 mℓ boiling 80% ethanol. The homogenate is centrifuged and the supernatant increased to 50 mℓ.

iv. Determination of Cyanide in Extract

Aliquots (0.1 to 0.2 mℓ) of the extract are pipetted into test tubes. The alcohol is removed by evaporation in a boiling water bath, and the tubes cooled to room temperature. Then 1 mℓ of 0.1 *M* phosphate buffer pH 6.0 and 0.1 mℓ enzyme are added and the tubes incubated at 30°C for 15 min. The reaction is stopped by adding 1 mℓ 0.2 *N* NaOH. One mℓ 0.2 *N* HCl is added to neutralize it and the cyanide content is estimated by addition of chloramine T and barbituric acid/pyridine reagent, as described earlier.

VII. CRUDE PROTEIN[9]

A. Kjeldhal Method

In the Kjeldhal nitrogen determination, the sample is digested in concentrated sulfuric acid. The sulfuric acid acts as a dehydrating and oxidizing agent. The nitrogen of the sample is transformed into ammonia. The ammonia is held back as ammonium in (NH^+_4) in the form of ammonium sulfate. Sodium hydroxide is then added to the solution, which transforms the ammonium ions into ammonia (NH_3) which is distilled, absorbed in a boric acid solution, and titrated with standard HCl.

Since the boiling point of the H_2SO_4 is not sufficiently high to oxidize organic substances quickly, K_2SO_4 is added to raise its boiling point. Also, $CuSO_4$ acts as a catalyst.

1. Reagents

1.   Sulfuric acid sp. gr. 1.84
2.   Catalyst mixture: grind together in a mortar 99.0 g of K_2SO_4 and 0.8 g of $CuSO_4$
3.   Sodium hydroxide (40%)
4.   Boric acid solution (Dissolve 10 g in 480 mℓ of boiling distilled water)
5.   Boric acid-indicator solution. (Add 20 mℓ of 0.1% bromocresol green in 95% alcohol and 4 mℓ of 0.1% methyl red to the above boric acid solution)
6.   Hydrochloric acid (0.1 *N* and 0.01 *N*)
7.   Digestion tubes and racks
8.   Kjeldhal distillation unit
9.   Microburette (5 mℓ)

2. Procedure

Weigh an appropriate amount of the sample into the digestion tube. Add 2 mℓ concentrated H_2SO_4 and 1 g of the catalyst mixture. Heat it until the solution becomes clear and colorless. A few drops of hydrogen peroxide may be added towards the end to assist in clearing.

The solution is cooled and transferred into the distilling tube. Arrange the delivery tube of the apparatus to be just below the surface of a 5 mℓ boric acid indicator solution in a 50 mℓ flask. Add 10 mℓ of 40% NaOH through the funnel and steam distilled until the volume of the liquid in the receiver reaches 10 mℓ. Wash the outside of the delivery tube with a little water and titrate the solution in the flask with 0.01 *N* HCl until the first pink tinge appears. Titrate a blank in the same way and obtain the titre value for the sample.

$$\% \text{ N} = \frac{\text{m}\ell \text{ HCl used} \times \text{Normality} \times 14 \times 100}{\text{mg sample}}$$

Most proteins contain about 16% Nitrogen, so that 16 mg N_2 = 100 mg protein 1 mg N_2 = 6.25 mg protein. The N value is multiplied by 6.25 to obtain the weight of the protein, i.e., % protein = % N × 6.25.

VIII. DETERMINATION OF CRUDE FIBER[10]

A. Reagents

1. Sulfuric acid — 1.25%. Weigh 1.25 g concentrated H_2SO_4 in a small beaker and increase with distilled water to 100 mℓ
2. Sodium hydroxide — 1.25%. Weigh 1.25 g NaOH and dissolve in distilled water and increase to 100 mℓ
3. Alcohol — 95%

B. Procedure

The sample is oven-dried at 105°C. Then 2 g of the powdered dried sample, is placed in a beaker and 200 mℓ of boiling 1.25% H_2SO_4 is added. Place the beaker on a hot plate and boil for 30 min, occasionally rotating the beaker. Cool and filter by suction through a Buchner. Rinse the beaker with two 50 mℓ portions of boiling water. Transfer the residue carefully to a beaker and add 200 mℓ, 1.25% NaOH. Boil for 30 min. Cool and filter as above and wash twice with 50 mℓ boiling water. Finally, wash with 25 mℓ 95% alcohol. Oven-dry the residue for 2 hr at 130°C. Cool in a desiccator and weigh. Heat 30 min at 600°C. Cool in a desiccator and weigh.

$$\% \text{ Crude fiber in the ground sample} = \frac{(\text{Loss in weight on ignition}) \times 10}{\text{Weight of sample}}$$

IX. PHENOLS[11]

The estimation of phenols with a Folin-Ciocalteau reagent is based on the reaction between phenols and phosphomolybdate, which results in the formation of a blue complex whose absorbance is measured at 660 nm.

A. Reagents

1. Folin-Ciocaltean reagent.
2. 20% Na_2CO_3

B. Procedure

Cut plant tissues into pieces of 1 to 2 cm, plunge into boiling alcohol and allow to boil for 5 to 10 min. Cool and crush tissues in a mortar with a pestle. Centrifuge and prepare the supernatant to a known volume.

Pipette 1 mℓ extract into a test tube, add 1 mℓ Folin-Ciocalteau reagent, followed by 2 mℓ Na_2CO_3 solution. Shake the tube and place in a boiling water bath for 1 min. Cool, dilute the blue solution to 25 mℓ with water, and measure the absorbance at 660 nm. Calculate the phenols in the sample from a standard curve prepared with catechol or caffeic acid.

X. CAROTENES

Carotenoids are present in some cassava varieties.[12]

A. Reagents

1. Acetone

2.   Petroleum ether
3.   Aluminium oxide
4. Na_2SO_4 (AR)

B. Procedure

Take 1 to 2 g fresh sample and grind it with 10 mℓ of a mixture consisting of acetone and petroleum ether (1:1) and containing 0.1% quinone. Allow it to settle and transfer the supernatant to a separating funnel. Grind tissue with fresh solution until no color can be extracted. The pooled filtrate in the separating funnel is rinsed with 50 mℓ portions of water and the water discarded. This is repeated twice. The solvent layer is then dried over any-hydrous Na_2SO_4 and concentrated under reduced pressure to a volume of 5 mℓ. Then 2 mℓ of the concentrated carotene extract is loaded on to a column of alumina (10 × 1 cm) containing 3% anhydrous Na_2SO_4. It is then eluted with 3% acetone in petroleum ether. Carotene is eluted first, leaving all other fractions in a zone at the top. The volume of this eluate is increased to the appropriate volume and the color is measured at 450 nm and the value calculated from a standard graph.

C. Preparation of Standard Carotene

A 50 mg sample of carotene is dissolved in a few milliliters of chloroform and increased to 100 mℓ with petroleum ether. Working standards of 0.2 to 4.0 mℓ are made by dilution with petroleum ether. The color is measured at 450 nm and a standard graph is prepared.

REFERENCES

1. **Aminoff, D., Binhley, W. W., Sheffer, R., and Mowry, R. W.,** Analytical methods for carbohydrates, in *The Carbohydrates,* Vol. 2, Pigman W. and Horton, D., Eds., Academic Press, New York, 1970, 763.
2. **Gilbert, L. M., Gilbert, G. A., and Spragg, S. P.,** Amylose and amylopectin from potato starch in *Methods in Carbohydrate Chemistry,* Vol. 4, Whistler, R. L., Ed., Academic Press, New York, 1964, 25.
3. **Soubhagya, C. M. and Bhattacharya, K. R.,** A simplified colorimetric method for determination of amylose content in rice, *Starke,* 23, 53, 1971.
4. **Ashwell, G.,** Colorimetric analysis of sugars, in *Methods in Enzymology,* Vol. 3, Colowick, S. P. and Kaplan, N. O., Eds., Academic Press, New York, 1957, 84.
5. **Somogyi, M.,** A new reagent for the determination of sugars, *J. Biol. Chem.,* 160, 61, 1945.
6. **Indira, P. and Sinha, S. K.,** Colorimetric method for determination of HCN in tubers and leaves of cassava (*Manihot esculenta* Crantz), *Ind. J., Agric. Sci.,* 39, 1021, 1969.
7. **Cooke, R. D.,** An enzymatic assay for the total cyanide content of cassava (*Manihot esculenta* Crantz), *J. Sci. Food Agric.,* 29, 345, 1976.
8. **Nambisan, B. and Sundaresan, S.,** Spectrophotometric determination of cyanoglucosides in cassava, *J. Assoc. Off. Anal. Chem.,* 67, 641, 1984.
9. Official Methods of Analysis, *Proteins,* Association of Official Analytical Chemists, Horowitz, W., Ed., 9th ed., Washington D.C., 1960, 169.
10. Official Methods of Analysis, *Crude Fibre,* Association of Official Analytical Chemists, Horowitz, W., Ed., 12th ed., Washington D.C., 1975, 136.
11. **Swain, T. and Hillis, W. E.,** The phenolic constituents of *Prunus domestica,* I. The quantitative analysis of phenolic constituents, *J. Sci. Food Agric.* 10, 63, 1959.
12. Official Methods of Analysis, *Carotenes,* Association of Official Analytical Chemists, Horowitz, W., Ed., 12th ed., Washington D.C., 1975, 821.

INDEX

Printed and bound by CPI Group (UK) Ltd, Croydon, CR0 4YY

22/10/2024

01777633-0011